Antoine Ruppaner

The Principles and Practice of Laryngoscopy and Rhinoscopy in Diseases of the Throat and Nasal Passages

Designed for the use of physicians and students

Antoine Ruppaner

The Principles and Practice of Laryngoscopy and Rhinoscopy in Diseases of the Throat and Nasal Passages
Designed for the use of physicians and students

ISBN/EAN: 9783337373627

Printed in Europe, USA, Canada, Australia, Japan

Cover: Foto ©berggeist007 / pixelio.de

More available books at **www.hansebooks.com**

THE

Principles and Practice

OF

Laryngoscopy & Rhinoscopy

IN

Diseases of the Throat and Nasal Passages.

Designed for the use of Physicians and Students.

With 59 Engravings on Wood.

By ANTOINE RUPPANER, M.D., M.A.,

Member of the American Medical Association; of the Massachusetts Medical Society; of the County Medical Society; of New York, etc.

NEW YORK:
A. SIMPSON & CO.

1868.

ENTERED according to Act of Congress, in the year 1867, by ANTOINE RUPPANER, M.D., in the Clerk's Office of the District Court of the United States for the Southern District of New York.

AGATHYNIAN PRESS, 60 Duane Street, N. Y.

PREFACE.

The value of the Laryngoscope and Rhinoscope in the diagnosis and treatment of disease beginning to be daily more appreciated by the profession, it is desirable that the principles and application of the same should come as speedily as possible into general use.

To advance this purpose, to enable the busy practitioner and the inexperienced student to overcome, by a moderate amount of practice and perseverance, that which is most essential to the successful application of this art, this treatise has been prepared, omitting nothing, as is hoped, which may be considered essential, but eschewing everything that has no immediate practical bearing upon the subject, tending only to confuse the student, and to increase the size and expense of the volume.

The author has consulted freely, in the preparation of these pages, the already extensive literature on this subject, and is particularly indebted to the works of Mackenzie, Gibb, Czermak, Von Brüns, Semeleder, and last but not least, to the labors and instruction of his former preceptors, Ludwig Türck, Doctor of Medicine, Professor at the Imperial University, Vienna, and Physician-in-Chief to the General Hospital, and Adelbert Tobold, Doctor of Medicine, Royal Counsellor of Health, and Docent at the University of Berlin, to both of whom he desires particularly to express his gratitude.

Having been commissioned during a visit to Vienna in 1865, by Prof. Türck, to translate and edit an English edition of his work (then in press, but since published), viz., "Die Krankheiten des Kehlkopfes und der Luftröhre, etc.," Wien, 1866, (The Diseases of the Larynx and Trachea, etc., Vienna, 1866), the preparation and execution of which English edition has been necessarily delayed for the present, on account of the great expense involved in the publication of such an extensive work with atlas, and its naturally limited sale, I have liberally availed myself of Prof. Türck's work, and especially of his admirably well-executed drawings, regarding this as an excellent opportunity to introduce an author of such high merit and reputation to an American medical public, hoping thereby to awaken an interest for the study of the original work among students in this department of practical science.

A task similar to the above had been assigned to me, in the fall of 1865, by my friend, Dr. A. Tobold, of Berlin, in regard to his work, then in press, viz., " Die Chronischen Kehlkopf's Krankheiten mit specieller Rücksicht auf Laryngoskopische Diagnostik und Locale Therapie," Berlin, 1866 (The Chronic Diseases of the Larynx, with special reference to Laryngoscopic Diagnosis and Local Therapeutics), the execution of which task for want of time and other reasons has been delayed only, but not abandoned.*

To Messrs. Otto & Reynders, the well-known instrument-makers, Chatham Street, New York, my acknowledgments are

---

* A translation of the above-mentioned work in English, by a self-appointed editor of Dr. Tobold's works having been announced a short time ago as being in press, I addressed a letter to Dr. Tobold, inquiring into the facts of the case, taking it for granted that no physician would presume to claim to be the editor of the work of a living medical author without first having obtained his personal consent thereto, thinking that I had been released from the otherwise agreeable duty of acting as translator and editor of this work, when I received at once an answer, from which I make the following extracts :

BERLIN, Sept. 7th, 1867.

*   *   *   *   "As I never have had any conversation or correspondence with any person except yourself in regard to the translation of my work entitled, "Chronic Diseases of the Larynx," etc., I beg you to make use of the following declaration whenever and wherever you may see fit to do so to protect both your and my own interests."

Declaration: "This is to certify, that I have authorized Dr. A. RUPPANER, of New York, solely and alone, to translate from the German into English, and to edit my work on the Chronic Diseases of the Larynx, as I consider him entitled and most competent to execute the same on account of his perfect knowledge of the German language, as well as owing to his thorough general medical education and his particular experience in the specialty of laryngoscopy."

DOCTOR A. TOBOLD.

due for supplying me with the greater part of the wood-cuts for this work.

Hoping that this effort may prove to be as useful as it is needed, and trusting that it may supply a want long felt by those interested in this branch of medical science, it is submitted to the kind and indulgent judgment of the profession.

FIFTH AVENUE HOTEL, NEW YORK,
      November, 1867.

TABLE OF CONTENTS.

PART FIRST.

CHAPTER I.
PAGE

SURVEY OF THE HISTORY, DISCOVERY AND DEVELOPMENT OF PRACTICAL LARYNGOSCOPY AND RHINOSCOPY - - 1

CHAPTER II.

THE PRACTICE OF LARYNGOSCOPY.

SECTION I. ESSENTIALS FOR LARYNGOSCOPIC EXAMINATION—THE LARYNGEAL MIRROR—THE REFLECTOR—LIGHT—MODUS OPERANDI—THE LAMP AND ITS POSITION - - 15

SECTION II. DIRECT LARYNGOSCOPIC EXAMINATIONS—POSITION OF PATIENT, PHYSICIAN, AND ARRANGEMENT OF APPARATUS—POSITION OF THE HEAD AND TONGUE OF THE PATIENT—MANNER OF HOLDING, INTRODUCTION, POSITION, AND CHANGES OF THE LARYNGEAL MIRROR—OBSTACLES ENCOUNTERED IN LARYNGOSCOPIC EXAMINATIONS—TONGUE—EPIGLOTTIS—SENSITIVENESS AND EXCITABILITY OF THE FAUCES—HYPERTROPHY OF TONSILS—IRREGULARITY OF RESPIRATION - - - 26

CHAPTER III.

AUTO-LARYNGOSCOPY - - - - - - - - 42

CHAPTER IV.

RECIPRO-LARYNGOSCOPY - - - - - - - 45

CHAPTER V.

INFRA-GLOTTIC LARYNGOSCOPY OR TRACHEOSCOPY - - 47

CHAPTER VI.

RHINOSCOPY—MODUS OPERANDI—SMALL MIRROR—REFLECTOR—TONGUE-SPATULA — PALATE-HOOK—PALATE-LASSO—CASES - - - - - - - - - - 49

CHAPTER VII.

CONCLUDING REMARKS ON THE PRECEDING CHAPTERS - - 60

PART SECOND.

CHAPTER I.

APPLICATION OF REMEDIES TO THE LARYNX AIDED BY THE LARYNGOSCOPE - - - - - - - - 64

CHAPTER II.

REMEDIES APPLIED—SOLUTIONS—POWDERS—SOLID CAUSTIC—ESCHAROTICS—HOW APPLIED—LARYNGEAL BRUSH—SPONGE-CARRIER—SYRINGE—PULVERISATEUR—CASES - 67

CHAPTER III.

ON OPERATIVE PROCEEDINGS WITHIN THE LARYNX.

SECTION I. INDICATIONS FOR OPERATIONS—SCARIFICATIONS AND OPENING OF ABSCESSES—CASES - - - - 103

SECTION II. THE REMOVAL OF MORBID GROWTHS FROM THE LARYNX—DECISION—EXCISION—PUNCTURE—CRUSHING—CAUTERIZATION—EVULSION—LIGATION—ECRASEUR—GALVANO-CAUTERY - - - - - - - - 111

CHAPTER IV.

THE APPLICATION OF GALVANISM TO THE LARYNX IN TEMPORARY OR PERMANENT APHONIA - - - - - 126

CHAPTER V.

GYMNASTIC OF THE LARYNX IN APHONIA—CASES - - - 133

CHAPTER VI.

REMOVAL OF FOREIGN BODIES FROM THE PHARYNX, LARYNX, AND TRACHEA—CASES - - - - - - 137

CHAPTER VII.

ANÆSTHESIA DURING OPERATIONS IN THE LARYNX - - 141

CHAPTER VIII.

CONCLUDING REMARKS - - - - - - - - 142

APPENDIX.

MIRROR OF NEUDÖRFER WITH FENESTRATED CANULA—LARYNGEAL SYRINGE, WITH MIRROR—SCHNITZLER'S SPRAY-PRODUCER—TÜRCK'S PORTE-CAUSTIQUE - - - - 145

MAGNIFYING INSTRUMENTS, MICROMETERS, AND DOUBLE MIRRORS - - - - - - - - - - 150

TABLE OF ILLUSTRATIONS.

		PAGE
Fig.	1.—Mirrors of different diameter (Türck)........	16
	2.—Profile of Türck's Mirror...................	17
	3.—Half Profile of the same....................	17
	4.—Türck's Reflector, supported by a Spring......	20
	5.—Türck's Independent Apparatus for Illumination	22
	6.—Tobold's Apparatus for Illumination..........	25
	7.—Position of the Hand and Mirror............	30
	8.—The parts of the Larynx seen in their natural position (Turck)........................	33
	9.—The parts of the Larynx as represented in the Laryngeal Mirror during examination (Türck)..	33
	10.—View of the Epiglottis, depressed (Türck).....	37
	11.—Omega-shaped Epiglottis (Türck)............	37
	12.—Border of the Epiglottis, depressed and rolled backward (Türck)........................	37
	13.—Epiglottis depressed, but partially raised to show the Arytenoid Cartilages and part of Vocal Cords (Türck)................................	37
	14.—Dr. Smyly's Recipro Laryngoscope...........	46
	15, 16, 17, 18, 19.—Türck's Tongue-Spatulas.....	51
	20.—Türck's Palate-Lasso.......................	52
	21.—Posterior view of the Naso-Pharyngeal space, after removal of the Dorsal Vertebræ of the anterior wall and a part of the side walls (Türck)	54
	22.—Rhinoscopic representation of the anterior and a part of the lateral walls of the Naso-Pharyngeal space (Türck)......................	55
	23.—Türck's Laryngeal Brush...................	70
	24.—Gibb's Laryngeal Brush.....................	70
	25.—Türck's Sponge-Carrier.....................	78
	26.—Türck's Sponge Syringe.....................	79
	27.—Gibb's Graduated Laryngeal Syringe..........	79
	28.—Lewin's Pulverisateur (Mackenzie)..........	84

TABLE OF ILLUSTRATIONS.

	PAGE
Fig. 29.—Gibb's Laryngeal Fluid Pulverizer	86
30.—Siegle-Bergson-Lewin's Inhaling Apparatus	87
31 (a).—Large Laryngeal Lancet (Türck's)	107
31 (b).—Handle to the above	107
32.—Tobold's Polypus Instrument for decision	114
33.—Tobold's Polypus Instruments for excision	114
34.—Tobold's Instrument for puncture	115
35.— " " "	115
36, 37.—Türck's large fenestrated Polypus Knives	116
38, 39, 40.—Türck's smaller fenestrated Polypus Knives	116
41, 42.—Türck's Knives without frame on one side	117
43, 44.—Türck's Sheath-knives	117
45.—Türck's Pincette	119
46.—Large doubly indentated Polypus-crusher (Türck)	119
47.—Large sharp Polypus-crusher (Türck)	119
48.—Small sharp Polypus-crusher (Türck)	120
49.—Sharp Polypus-crusher, with blades placed square (Türck)	120
50.—Single indentated Polypus-crusher (Türck)	120
51.—Large double indentated Polypus-crusher (Türck)	120
52.—Gibb's Laryngeal Ecraseur	121
53.—Mackenzie's Laryngeal Galvanizer	129
54.—Türck's Laryngeal Pincers	140
55.—The same in detail	140

APPENDIX.

56.—Neudörfer's Mirror for Infra-glottic Laryngoscopy or Tracheoscopy	145
57.—Tobold's Laryngeal Syringe with Mirror attached	147
58.—Schnitzler's Spray-producer	148
59.—Türck's Porte-caustique	149

LITERATURE.

Works which have been consulted, bearing upon the subject of Laryngoscopy, Rhinoscopy and Inhalation. In this list are not included the numerous articles on these subjects scattered through the periodical medical literature, both foreign and American.

Czermak.—" Der Kehlkopfspiegel und seine Verwerthung für Physiologie und Medizin." (The Laryngoscope, its application to Physiology and Medicine). Second Edition, Leipzig, 1863. The first edition, translated from the French, is published by the New Sydenham Society. Vol. XI.

Türck.—" Beiträge zur Laryngoskopie und Rhinoskopie." (Contributions to Laryngoscopy and Rhinoscopy). Vienna, 1860. " Notizen zur Rhinoskopie." (Notes upon Rhinoscopy). Vienna, 1860. " Praktische Einleitung zur Laryngoskopie." Wien, 1860. (Practical Introduction to Laryngoscopy); also, " Récherches Cliniques sur diverses Maladies du Larynx." Paris, 1862; also, "Clinical Researches on different Diseases of the Larynx," etc. (English translation from the French). London, 1862, Williams and Norgate; finally, " Klinik der Krankheiten des Kehlkopfes und der Luftröhre," etc. (Clinic of the Diseases of the Larynx and Trachea, etc). Vienna, 1866; also, " Atlas zur Klinik der Kehlkopfkrankheiten, in 24 Chromo-lithographirten Tafeln." (Atlas to the Clinic of the Diseases of the Larynx and Trachea, of 24 Chromo-lithographic Plates). Vienna, 1866.

Semeleder.—" Die Rhinoskopie und ihr Werth für die Aerztliche Praxis." Leipzig, 1862; also, " Die Laryngoskopie, und ihre Verwerthung für die Aertztliche Praxis." Wien, 1864. (Rhinoscopy and Laryngoscopy, their

Value in Practical Medicine, both Monographs combined, by Dr. E. T. Caswell, Providence). New York, 1866.

Voltolini.—" Eine Monographische Arbeit zur Fünfzigjährigen Jubelfeier der Universität Breslau, Aug. 1861." (Monograph upon the Occasion of the Semi-centenarian Jubilee of the University of Breslau, Aug. 3, 1861), and various papers by that author in " Virchow's Archiv.," " Jahrbuch der Gesellschaft der Aerzte zu Wien," " Deutsche Klinik," etc. (1860 and 1861).

Störk.—" Rhinoskopie." Journal of the Society of Physicians of Vienna, No. 26, 1860.

Dauscher.—Journal of the Society of Physicians of Vienna, No. 38, 1860.

Zsigmondy.—" Neue Folge Galvanokautischer Operationen." (New Results of Galvano-caustic Operations). " Oesterr Zeitsch. f. Prakt. Heilkunde." No. 39, 1860. (Austrian Journal of Practical Medicine).

Gerhardt u. Roth.—" Ueber Syphil. Krankheiten des Kehlkopfs." (Syphilitic Diseases of the Larynx). Virchow's Archiv., Bd. XXI.

Lewin.—" Die Laryngoskopie, Beiträge zu ihrer Verwerthung für Praktische Medizin." (Laryngoscopy, a Contribution to its Value in Practical Medicine). Berlin, Hirschwald, 1860 ; " Klinik der Krankheiten des Kehlkopfes." I Band, etc. (Clinic of the Diseases of the Larynx, Vol. I). " Treatment of the Diseases of the Respiratory Organs by Inhalation, with Special Reference to those Diseases of the Larynx made known by the Laryngoscope." With 25 wood-cuts. Second enlarged and revised edition. Berlin, 1865.

Moura-Bourouillou.—" Cours complet de Laryngoscopie." De la Haye, Paris, 1861. " Traité Pratique de Laryngoscopie et de Rhinoscopie, suivi d'observations." A. De la Haye, Paris, 1865.

Battaille.—" Nouvelles Récherches sur la Phonation." Paris, V. Masson, 1861.

Fauvel.—" Du Laryngoscope au point de vue Pratique. Paris, V. Masson, 1861.

JAMES.—" Sore Throat; its Nature, Varieties, and Treatment, including the use of the Laryngoscope as an aid to Diagnosis." London, Churchill, 1861.

MERKEL.—" Die Funktionen des Schlund und Kehlkopfes." (The Functions of the Pharynx and Larynx). Leipzig, O. Wigand, 1862.

YEARSLEY.—" Introduction to the Art of Laryngoscopy." London, Churchill, 1862.

WAGNER, OF NEW-YORK.—" Zur Laryngoskopie und Rhinoskopie." (Laryngoscopy and Rhinoscopy). Oesterr. Zeitch. f. Prakt. Heilkunde. No. 6, 1862.

TOBOLD.—" Lehrbuch der Laryngoskopie und des Localtherapeutischen Verfahrens bei Kehlkopfkrankheiten." (Manual of Laryngoscopy and of Local Therapeutics in Diseases of the Larynx). Berlin, Hirschwald, 1863; also, "Die Chronischen Kehlkopfskrankheiten, mit specieller Rücksicht auf Laryngoskopische Diagnostik und Locale Therapie." (The Chronic Diseases of the Larynx, with Special Reference to Laryngoscopic Diagnosis and Local Therapeutics). Berlin, Hirschwald, 1866.

BRÜNS, VON.—" Die erste Ausrottung eines Polypen in der Kehlkopfshöhle durch Zerschneiden." (The First Extirpation of a Polyp in the Cavity of the Larynx by Cutting). Tübingen, Laupp, 1862. Supplement to the same, do., 1863. " Die Laryngoskopie and Laryngoskopische Chirurgie." (The Laryngoscope and Laryngoscopic Surgery). Tübingen, 1865. Atlas to the same, fol., maps, 8 plates, 2 colored.

SIEVEKING.—" Practical Remarks on Laryngeal Disease." London, 1862.

GIBB.—" The Laryngoscope. Illustrations of its Practical Application, and Description of its Mechanism." London, Churchill, 1863. " On Diseases of the Throat and Windpipe, as reflected by the Laryngoscope; a Complete Manual upon their Diagnosis and Treatment." London, Churchill, 1864.

RUSSELL.—"On Laryngeal Disease," etc. London, 1864.

WALKER.—" On the Laryngoscope." London, Richards, 1866.

JOHNSON.—"Two Lectures on the Laryngoscope, and Directions for its Use." London, Hardwicke, 1864.

GUILLAUME.—" Essai sur la Laryngoscopie et la Rhinoscopie."
Paris, De la Haye, 1864.

BAUMGARTEN, DR. J.—" Die Krankheiten des Kehlkopfes und
deren Behandlung," etc. (The Diseases of the Larynx
and their Treatment, together with a Notice of some
New Inhalation Apparatus, and an Introduction to Laryngoscopic Examination), with wood-cuts. Freiburg, 1865.

MACKENZIE, MORRELL.—"The Use of the Laryngoscope in
Diseases of the Throat, with an Appendix on Rhinoscopy." London, 1865.

DIXON, THOMAS.—" On Diseases of the Throat; their New
Treatment by the Aid of the Laryngoscope." London,
Renshaw, 1865.

RUHLE.—"Clinic of the Diseases of the Larynx." Berlin,
Hirschwald, 1865.

WALDENBURG.—" Inhalation of Pulverized Fluids, Steam and
Gas, and their Effect in Diseases of the Respiratory Organs." Berlin, Reimer, 1864.

BEIGEL.—" On Inhalation as a Means of Local Treatment of
the Organs of Respiration by Atomized Fluids and
Gases." London, 1866.

COHEN.—" Inhalation: its Therapeutics and Practice." Philadelphia, Lindsay & Blakiston, 1867.

COSTA, DA.—"Inhalations in the Treatment of Diseases of the
Respiratory Passages, particularly as Effected by the Use
of Atomized Fluids." Philadelphia, Lippincott & Co.,
1867.

THE PRINCIPLES AND PRACTICE

OF

LARYNGOSCOPY

AND

RHINOSCOPY.

PART FIRST.

Chapter I.

SURVEY OF THE HISTORY, DISCOVERY AND DEVELOPMENT OF PRACTICAL LARYNGOSCOPY AND RHINOSCOPY.

From the beginning of the present century the efforts of physicians have been chiefly directed to the correct diagnosis of diseases.

Brilliant as have been their achievements, still the correct diagnosis of the diseases of the larynx as well as those of the components parts of the naso-pharyngeal cavity remained, if not impossible, yet obscure, by reason of these parts being hidden from view.

To recognize the changes in the pharynx, larynx and posterior nasal cavities, it was customary to cause the patient to open the mouth, and depress the tongue at the same time, by means of a spatula, when the velum palati and its pillars, the pharynx, and sometimes even the epiglottis were brought within sight; but for the examination of the interior of the larynx this remained insufficient.

Efforts have not been wanting to devise means or to construct instruments, whereby the eye of the physician might be enabled to unveil that which was hidden. If we except Bozzini,* whose experiments were not directly applied to the larynx, but who described an instrument whereby to throw light into any cavity of the human body accessible from without, Senn,† of Geneva, appears first in 1827 to have conceived the idea, and to have taken the first practical steps towards examining the interior of the larynx, by means of a small mirror introduced into the mouth.

Then followed Dr. B. C. Babington,‡ in 1829, March 18th, who laid before the Hunterian Society of London, his "Glottiscope," a small oblong piece of looking-glass, set in wire, with a long shank. The reflecting portion was placed against the palate, whilst the tongue was held down by a spatula, when the epiglottis and the upper part of the larynx became visible in the glass.

---

* P. H. Bozzini, Weimar, 1807, and Hufeland's Journal, 1806. Vol. XXIV, p. 107.

† Gazette Hebdomadaire de Médecine et de Chirurgie. Paris: 1863. Page 263.

‡ The London Medical Gazette. London: 1829. Vol. iii., p. 555.

Dr. Babington afterwards had his mirror made of polished steel, and in one he combined a tongue depressor with it. He is also reported to have used oval shaped mirrors, convenient in case of enlarged tonsils; and to have illuminated the throat by reflecting the light of the sun from a mirror held in the left hand.

Trousseau & Belloc* in 1837, refer to a laryngeal speculum which was made for them by a Mons. Sellique, an ingenious mechanic, who had himself been subject to laryngeal phthisis. This instrument which consisted of two tubes, through one of which light was thrown on the glottis, whilst through the other the image of the glottis was reflected from a mirror placed at its gutteral extremity, was, however, difficult of application, since not one in ten could bear its presence in the throat. Soon it was declared void of practical utility and abandoned. Although Bennati† is said to have been enabled by it to obtain a view of the glottis, which statement however is strongly doubted, as Bennati,‡ then Physician-in-Chief to the Italian Opera in Paris, does not allude to the fact in the least in his writings on this subject.

Baumes of Lyons§ about 1838, laid before the Medical Society of that city, a laryngeal speculum of a somewhat complicated construction, but about the

---

* Traité pratique de la phthisie laryngée, de la laryngite chronique, et les maladies de la voix, par Trousseau et Belloc, Paris: 1837, p. 179.

† Opus cit: Trousseau et Belloc. p. 180.

‡ Etudes physiologiques et pathologiques sur les organes de la voix humaine, Paris: 1833 and 1834.

§ Compte rendu des travaux de la Société de Médecine de Lyons: 1840, p. 62.

practical utility and the results of its application no record has been published.

In 1840, Liston, in his *Practical Surgery** mentions the successful employment of this instrument in medical practice in the following words, page 417, "Ulcerated glottis: a view of the parts may be sometimes obtained by means of a speculum; such a glass as is used by dentists, on a long stalk previously dipped in hot water, introduced with its reflecting surface downwards, and carried well into the fauces."

Beyond a doubt it is established, that the late Dr. Warden, in the year 1844, employed a powerful Argand lamp before which he placed a prism of flint-glass, by means of which contrivance he succeeded in seeing disease of the glottis in two cases.† These are so accurately described by him in all their details, that Warden must be regarded as the first observer who had by artificial light examined an unhealthy larynx in the living.

Dr. Avery, of London, is also reported to have worked successfully, from 1844, till his death in 1854 cut short his investigations, in the construction of a laryngoscope and other instruments for the examination of internal organs, but he has published nothing on the subject.‡

Thus matters stood, when in the year 1855, the Spaniard Manuel Garcia, a singing teacher in London,

---

* Practical Surgery, with 50 engravings on wood; by Robert Liston, Esq.

† London Medical Gazette. New series, 1844, vol. II., page 256. And British and Foreign Medico-Chirurgical Review, January, 1863

‡ Introduction to the Art of Laryngoscopy, by Dr. Yearsley, 1862.

published his observations on the formation of the human voice, in an interesting paper, entitled "*Physiological Observations on the Human Voice*," in the Proceedings of the Royal Society.*

Garcia was the first, then, to employ the laryngeal mirror for physiological purposes; the first who succeeded in obtaining a view of his own larynx, so as to describe accurately the parts he saw in the mirror and their respective motions. Having long studied the anatomy and physiology of the larynx as the organ of voice, a great desire to see the movements must have been the natural consequence. How he at length obtained the desired object he describes himself. Standing with his back to the sun, he held a looking-glass in his left hand before his face; the sun's rays were thus reflected by the glass into his open mouth. Then he introduced a dentist's mirror—previously warmed—into the back of his mouth, and thus he saw the reflection of his larynx in the looking-glass. Auto-laryngoscopy was discovered; it was an indisputable fact, and to Garcia belongs the credit of having demonstrated it. His method is to-day the simplest and best for auto-laryngoscopic examination.

Strange to say, as Senn's, Babington's and Warden's observations were sown without fruit; as Liston's teachings were forgotten, so Garcia's brilliant results were received with distrust, nay, positively doubted and rejected. Still, in spite of the continued opposition of those who, either from old age, pride, caste, or ignor-

---

* Observations on the Human Voice, Philosophical Magazine and Journal of Science, vol. X. and Gazette Hebdom. de Méd. et de Chir. 16 Nov., 1855.

ance, oppose all progress in experimental or practical science, the seed that had been sown by Garcia came to be the germ of many important results and discoveries.

Dr. Ludwig Türck, Professor at the University of Vienna, and one of the Physicians-in-Chief to the general hospital, ignorant at the time, as he distinctly states, of the experiments and writings of Garcia,* had, after a long series of most varied experiments, made during the summer of 1857 on dead subjects and on patients under his care, in the general hospital, succeeded in giving the laryngoscope so convenient a shape as suited it best for the purpose of examination both of the larynx and adjoining parts in a great number of individuals.†

When demonstrating to Professor Ludwig in the summer of 1857 the interior of the larynx on a patient in his ward, Türck was for the first time made aware of Garcia's researches.‡ Garcia's investigations, however, were purely physiological; Türck, on the other hand, was seeking for larger results, viz, means for correct diagnosis from which would spring a more rational

---

* Türck, Clinic of the diseases of the Larynx and Trachea, Vienna, 1866, page 2 : I desire here explicitly to state, notwithstanding the assertion to the contrary in many of the works published in Europe on this subject, from my knowledge of the facts in the case, and from my acquaintance with Professor Türck as a man of veracity, that he was actively engaged in laryngoscopic researches before he had ever heard of Garcia's experiments, or before he ever had read Garcia's articles on the subject published in the Transactions of the Royal Society.—[AUTHOR.]

† Report of the Imp. R. Society of Physicians of Vienna; Meeting of April, 1858. In the year of 1857, Prof. Türck treated, in seven wards of his division of the hospital, 1,873 patients, among which there were only 275 cases of the nervous system. The greater number of the remainder suffered from diseases of the throat and chest.— *Ibid. Clinical Researches*; London, 1862.

‡ Op. cit. page 2.

and successful treatment of the various affections of the throat. Türck used direct sunlight for his demonstrations, in the absence of which, during a part of the winter, from the hospital wards, his experiments were discontinued.

Towards the end of that same year (1857) Dr. Czermak, Professor of Physiology at Krakau, but at that time sojourning in Vienna, who had heard of the flattering results of Professor Türck's investigations, borrowed the latter's mirrors for the express purpose of repeating Garcia's physiological experiments and for further investigations.*

Whilst Czermak repeated and verified Garcia's experiments with Türck's mirrors, he, Türck, did not remain idle, but improved upon his mirror until he brought it to its present simple perfection—a mirror now universally in use. During the clear days of the latter part of winter, he resumed also his examinations on patients in the wards, and the result answered his most sanguine expectations.

Again, on the 16th day of March, 1858, Czermak accompanied by University Docent and Secundar Physician, Dr. Gruber, visited Professor Türck, and sought and obtained permission to use the borrowed mirror in the syphilitic wards of the hospital. It was not many days† before it became evident Czermak was deeply impresed with, and perceived the practical value of the instrument, which became so potent in his hands, and for which he made most strenuous efforts to secure its

---

† Op. cit. page 203.   † Wiener Med. Wochb.

recognition by the whole civilized world. Setting to work with great zeal and energy with the mirrors of Professor Türck, he first sought to render us independent of sun-light by employing artificial light, using for that purpose the large reflector of the ophthalmoscope, thus rendering the laryngoscope available at all times for practical purposes. In his first article on this subject* Czermak called attention to the fact, that by means of the laryngeal mirror the eye becomes the sure guide of the hand. Further, he asserted the possibility of inspecting (in a manner similar to the larynx) the posterior surface of the soft palate, the posterior openings of the nasal fossæ and the superior parts of the pharynx, by simply turning the little mirror upwards and by raising the velum up as far as practicable.

This was the advent of the kindred art of Rhinoscopy.

The foregoing pages sufficiently establish the fact, that for the discovery of the art of laryngoscopy and its kindred rhinoscopy, as well as for the progressive steps of improvement and development which have led to their adoption in the diagnosis and treatment of disease we are chiefly indebted to Türck and Czermak.

The gradual development of this subject is illustrated by the numerous, interesting, scientific, and instructive publications of these two gentlemen. Every learned reader must, however, regret that there should have arisen a controversy, on the whole uninteresting and unprofitable to the mass of the medical profession, as to the claim of priority in the art of Laryngoscopy.

---

* Op. cit.

Where both have fought well and gained renown, when success pours into the lap of both its reward, who shall decide between the two?

As an act of justice, however, and for the benefit of those who have neither time nor inclination to read all the published evidence, pro and con, of the case, the controversy shall receive a passing notice.*

To prove priority we have to answer:

1st. Who conceived the idea and made use of the laryngeal mirror first?

2d. Who brought it to its present simple state of perfection?

3d. Who demonstrated first its usefulness in the diagnosis and treatment of disease by actual experiment?

4th. Who demonstrated first the art of rhinoscopy?

5th. Who labored most to compel for this art the acknowledgment due it from the medical world?

If we exempt, as we justly may, former and imperfect attempts at laryngeal examination, and acknowledge Garcia's claim to originality in the matter of auto-laryngoscopy, being obviously quite distinct and indisputable; the contest is narrowed down to Czermak and Türck. Türck deliberately states,† and this is the first printed reference we have on this vexed question—that being ignorant of the work of his predecessor, Garcia, he happened accidentally upon the idea early in 1857 (if not before) to employ a small mirror in the examin-

---

* We draw our inferences from the works of the two persons most interested, viz, Türck and Czermak.

† Op. cit. Page 2, line 16.

ation of diseases of the larynx. Professor Ludwig, in the summer of 1857, first informed him of Garcia's experiments. Great were the difficulties, but, in spite of obstacles, in many cases he accomplished his object. Winter interrupted his experiments for a time. Then (winter 1857) Czermak borrowed Türck's mirrors, and in March, 1858, obtained permission to make use of the identical mirrors in the syphilitic wards of the hospital.* Did Türck remain idle during the latter part of that winter and the first months of 1858? No! he continued to improve on his original mirror till it reached a perfection then which nothing better has to this day been produced in spite of various tinkerings with the same—such as making the mirrors square, then with round edges, oblong and oval, and lastly with a convex border above and a concave one below. One and all, so many attempts to make the most simple, practical instrument complicated, they have failed.

March 27, 1858, Czermak published his first paper,† in which he urgently recommended the laryngeal mirror to the medical profession for practical experiments. Up to this date Türck had published as yet nothing on the subject, but at a first meeting of the Medical Society of Vienna, April 9, 1858, (the first after Czermak's article appeared), Türck deliberately and openly claimed priority in the matter, which Czermak not only did not dispute, but published a second

---

* Op. cit. Page 3.
† Op. cit., page 3. Czermak Wiener Med. Wochenschrift, No. 13, March 27, 1853.

article,* in which he, in the clearest manner possible, acknowledged Türck's claim to priority. How he could do otherwise we cannot conceive. What he accomplished in this new field of discovery at first, he brought about with Türck's instruments, which he borrowed, and which were the result of Türck's ingenuity and study. The one furnished the other with *the thing* already discovered, necessary to accomplish *another thing* dependent wholly upon the first for its accomplishment; for we look upon the laryngeal mirror as the laryngoscope. Without it, laryngoscopy falls to the ground. Would the inventor of a machine, who delivers or lends his invention to another friend for a short time to repeat the same experiments for which the original machine is intended, (should the latter succeed in the use of it,) would the inventor of that machine therefore, we ask, be rightfully regarded as the inventor? or would he be considered as having relinquished his claim to his friend who had borrowed the machine; although the latter may even have added improvements, leaving, however, the original principle of construction by which the utility of the original machine may have been increased?

So far, Türck's claim to priority is fully and incontrovertibly established by facts which ought to satisfy the inquirer, whatever sophistry may have been employed, or may hereafter be used, to explain these facts away. Nor will—aside from the claim of priority —the labors of Prof. Türck in this department permit

---

* Wiener Med. Wochenschrift, No. 16, Extra April 17, 1858.

him to remain in the background. His late great, learned and superb work on the *Diseases of the Larynx and Trachea** places him in the very foremost position in this department of science. It will ever be pointed out as one of the best German classics in medical literature. It will be long before it will have a rival in its own or kindred languages. To such genius it is a pleasure and privilege to do homage and to render our acknowledgments.

The Commissioners of the Institute Imperial of France, (section of medicine and surgery), at the concourse for the prize Montyon, March, 1861, unwilling to enter into a discussion as to the relative merits of priority of the two competitors, Türck and Czermak, but being satisfied that the researches of both had equally contributed to render laryngoscopy a useful art, susceptible of good service in the diagnosis of the diseases of the pharynx and larynx, granted to both of the competitors an honorable mention, accompanied by a prize of 1,200 francs each, placing Türck's name first.

Thus having claimed justice for Professor Türck, we shall not overlook Czermak's merits. He is entitled to our praise; it is hard earned. Discussing the rival claims of discoverers, Sydney Smith is reported as having said: "That man is not the first discoverer of any art who first says the thing, but he who says it so long and so loud and so clearly that he compels mankind to hear him—the man who is so deeply impressed with the importance of his discovery that he will take

---

* Wien, 1866.

no denial, but at the risk of fortune and fame, pushes through all opposition, and is determined not to perish for a want of a fair trial." On grounds like these, Czermak has the strongest indisputable claims for our recognition of his merits. He first rendered artificial light available in our art. It was he who reduced the art to a system ; he who, from the very first, insisted and laid stress upon the importance of the laryngoscope in the diagnosis and the treatment of laryngeal diseases, and illustrated the correctness of his sayings by practical results in practice. To him, also, are we indebted for the art of Rhinoscopy. Czermak, finally it was, who labored much at home and in foreign lands, till he became the great teacher of the arts of Laryngoscopy and Rhinoscopy, as he had been one of their principal improvers.

Semeleder, Voltolini, Lewin, Störk, Ruete, Merkel, Traube, Middeldropf, in Germany, Mandl, Maura-Bourouillon, Bataille, Cusco, and Fauvel, in France, Mackenzie, Gibb, and Johnson, in England, received their first impulses from him, having witnessed his demonstrations.

The result of their many important investigations, in this department, is recorded in our medical literature. Surely, the claim to priority may be dismissed, without feeling, by one whose name is so indissolubly, so praiseworthy, so honorably interwoven with this subject.

This historical survey shows that the discovery and development of the arts of Laryngoscopy and Rhinoscopy, and all real and permanent progress made in the same, dates only from the beginning of 1857. Ten years have sufficed to develope this branch of science to a perfection, so as to conquer for it the recognition and support of the whole scientific world. As late as 1857, Mr. Porter, an English surgeon, wrote: "How is a man of experience, when he meets with a case of laryngeal disease, to know whether it is caused by an œdematous condition of the submucous tissue—by a chronic thickening of the mucous membrane itself—by laryngeal ulceration—by destruction of the cartilages—by the presence of abscess, or tumor, or by any other of those numerous affections which dissection so frequently shows us to be the occasion of death?" And he suggests that: "Perhaps, by reason of the difficulty of the subject, it will be long before we possess the same accuracy of information with reference to the affections of the windpipe that has been attained in other diseases."*

What a change has been effected since Mr. Porter wrote the above, and that change has been brought about, as before said, in ten years! The larynx has ceased to be an organ hidden from sight; it is brought within the range of our vision as much as the eye is, and more so than the ear. Dr. Johnson's answer to Mr. Porter's question, fittingly, may close this discussion. "The man of experience has now only to *look* into the larynx, and he will *see* what is the form of the disease with which he has to deal."

---

* Quoted by Dr. Johnson in his Lectures on the Laryngoscope, 1864.

CHAPTER II.

THE PRACTICE OF LARYNGOSCOPY.

SECTION I.

a. Essentials for Laryngoscopic Examination.

THE essentials for the purpose of undertaking a laryngoscopic examination are : a source of light from which the rays are thrown either directly, or as is usual indirectly, into the throat by means of a mirror—a reflector—from whence these rays, by the use of the small mirror (laryngeal mirror) placed in a position obliquely beneath the palate, are cast into the cavity of the larynx to illuminate the latter in such a manner that its image is reflected in the laryngeal mirror, and seen by the observer.*

a. The Laryngeal Mirror—Laryngoscope—Kehlkopfrachenspiegel of Prof. Türck, is simply a small looking-glass, " whose end is to hold, as 'twere, the mirror up to nature."†

It is, as it were, the only armament we absolutely need to examine the larynx by direct sunlight without reflection. What a store of knowledge this little instrument has revealed to us ! How instructive its

---

* Victor von Brun's Laryngoscopische Chirurgie. Tübingen, 1865.
† Quoted by Dr. Johnson.

lessons! What benefits it enables us to bestow upon our suffering fellow beings! This small mirror, either manufactured of the best polished steel, or of glass, set in a silver frame, is fixed into a stem of metal, and this into a handle of sufficient length, so that the insertion of the stem into the handle will rest upon the first phalanx of the middle finger, and the extremity of the handle is supported by the soft parts forming the angle of union between the thumb and index finger. Mirrors intended for operative purposes must have longer handles The stem ought to be inserted firmly into the mirror at an angle of 120°–125° as first pointed out by Türck,* whose round mirror is now universally in use, and superior to any other known. Mirrors vary also in size, and such choice must be made as is best adapted to the condition of the person to be examined. Gradually a tolerably large one can be introduced, where at first the smallest size was hardly tolerated. Finely silvered glass mirrors are preferable to those made of steel or other metal; the latter soon loose their polish, quickly cool, and become dimned by the breath of the patient.

FIGURE I.

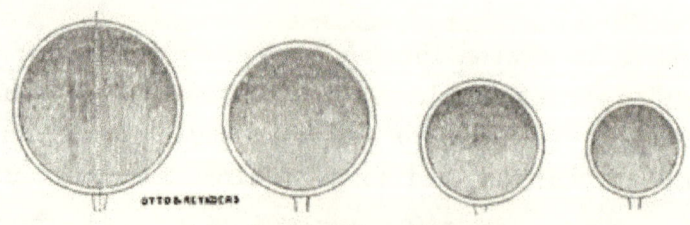

Mirrors of different diameter.

---

* Op. Cit. Page 4.

FIG. 2.   FIG. 3.

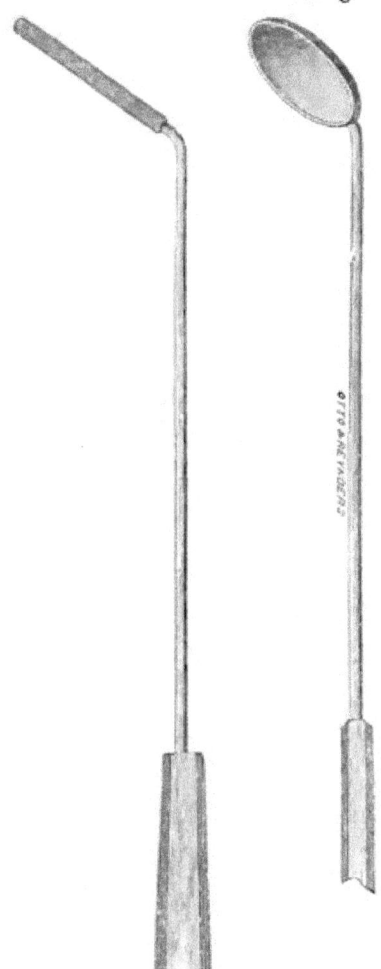

Profile of Türck's Mirror, and half profile of the same.
The handle is not given in full length.

b. The reflector is a round concave mirror of varying diameter, fastened to the forehead or placed in front of one eye of the operator, if perforated, for the purpose of reflecting the light of the sun, or of a lamp, into the throat. Czermak, Semeleder, Von Bruns and others have each adopted and recommended different ways of handling the reflector during examination. Every operator has his preference. Where, for want of sun-light, direct examinations are rarely made, and these in haste, the reflector placed on the forehead—rather than in the front of the eye—will soon be appreciated as both convenient and advantageous. Examinations made with the reflector, in the first mentioned position, are more readily executed, by beginners in the art especially, and are as satisfactory in their results as when the mirror is placed before either eye, which requires special training, without yielding superior results. In stationary apparatus, such as Czermak's, Van Bruns, Tobold's, Lewin's, and others, the reflector is attached to the illuminating apparatus itself. When used with sunlight, the patient is placed with his back to the sun and the operator sits opposite him. In case a lamp is employed, it is to be placed to the right side of the patient's head, so that the flame is about on a level and a little behind his right ear. By good sunlight, with a laryngeal mirror and a reflector a perfect examination of the larynx can be made without further addenda.*

c. Light, which is either natural or artificial, is a third essential in laryngoscopy. Sunlight can be supplied for this purpose either without or with reflectors. In the first position, without reflector, which is the most simple, the patient faces the light whilst the operator introduces the mirror into the mouth and places it obliquely against the palate. In the second position, already described with reflectors, the patient is placed with his back to the sun and light guided into the throat by means of the reflector.

There can be no question but that the sun gives the clearest and purest light—more desirable for our purpose than any other—provided we could have its benefit at all times and in all places, could ward off the great heat from the patient's head, which a lengthy examination renders insupportable, and could stay his course for the time, so as to abolish the necessity, both for patient and

---

*Professor Türck's reflector, with spring-supporter, seemingly belonging here, will be spoken of hereafter under caput "light."

doctor, for changing position every few minutes, for the purpose of keeping pace with his constant progress. Dependent then as we are upon the sun's favor, and rarely—especially within our city walls—the recipients of his benefactions, we are mostly, if not entirely, obliged to rely upon artificial light in the application of our art.

The simplest contrivance in this respect for our purpose is the German student's lamp, or an Argand-gas burner, where the luxury of gas is enjoyed. Of the lamp and its position we have already spoken, which, with the reflector is all that is needed to make illumination perfect.

Our description of the reflector (on page 17) would be incomplete, did we not here introduce Prof. Türck's simple, ingenious, and eminently practical combination of both reflector and supporter.

With this reflector on the forehead, its position capable of being changed in an instant, both hands of the operator remain free to depress the tongue, to apply caustic or other local remedies, or to perform various operations. In cases of diptheria and scarlatina especially, an examination can be made in a more thorough manner, with less loss of time than in the ordinary way, and without obliging the patient to raise the head from the pillow. There being more or ess risk of inhaling the patient's breath under such circumstances, and of having some morbid secretions coughed into the face, the physician is thus less exposed to infection than otherwise.

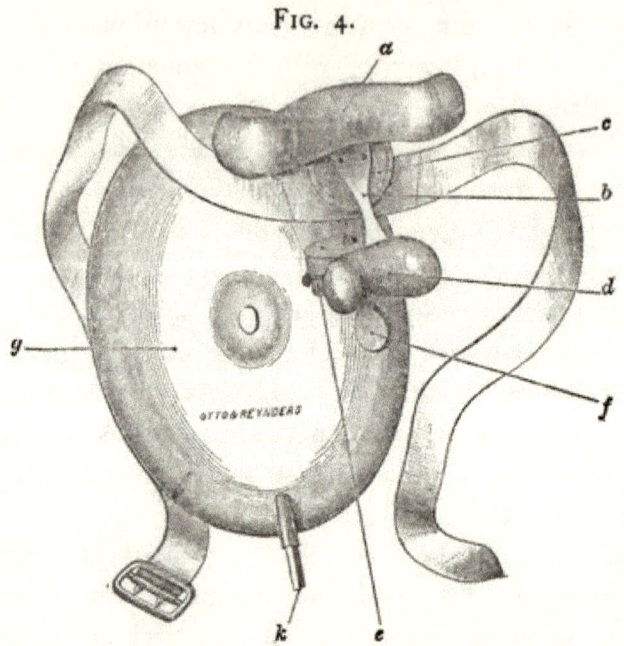

Fig. 4.

Türck's Reflector supported by a Spring.

a. Steel-spring projecting at both ends, upholstered with flannel and covered with black silk.

b. A narrower, quite strong somewhat bent steel-spring, bearing considerable resistance, inserted at right angles into spring *a*.

c. A round strong metal plate at the point of union, to the anterior side of which is attached an elastic band with buckle, which serves as head band.

d. Square upholstered saddle, attached to the lower, broader extremity of the descending spring, (*b*) which is to be placed at the apex of the nose.

e. Socket joint at the anterior surface of the saddle into which is inserted the ball attached to the posterior surface of the reflector.

f. Screw to render the joint more or less firm, or to remove the mirror altogether.

g. Reflector of $3\frac{1}{2}$-4 inches diameter taken from Türck's apparatus for illumination.

k. The handle can be used to change the position of the mirror.

This supporter will readily adapt itself to the forehead, is very light, and presses therefore, not uncomfortably upon the root of the nose, a great objection to all similar contrivances.

Rotation is easily accomplished, and the reflector

placed readily in the middle above the eyes, or replaced before either of them.

To Professor Türck also belongs the credit of having first used for illumination in laryngoscopic examinations the so-called Schuster-Kugel, (Shoemaker's globe). Both globes and lenses have been used without reflectors.

Moura's Pharyngoscope differs from the above simply in the fact, that a lens is substituted in the apparatus for the glass globe.

A simple, compact apparatus appeared in France some two years ago, but not generally in use, in which a metallic reflector is placed posteriorly to the light, and a bull's-eye lens or condensor anteriorly to the same, giving a good volume of light, besides being easily managed.

By fastening a second mirror to the bull's-eye lens, the apparatus answers perfectly for auto-laryngoscopic purposes. In examinations at the bedside, where gas is not to be had, we are able, if we can obtain simply a candle or a common lamp, to accomplish all we desire with it.

To speak of all the instruments which have been constructed to direct and intensify the volume of light such as Czermak's, Lewin's, Von Bruns', Winterich's, and those of several others not less active experimenters, would occupy too much space. We shall only describe two, those of Türck and Tobold, which are in daily requisition in our operating room, the practical merits of which in daily use for a long time, we have fully learned to appreciate above all others that have been tried from time to time by us.

FIG. 5.

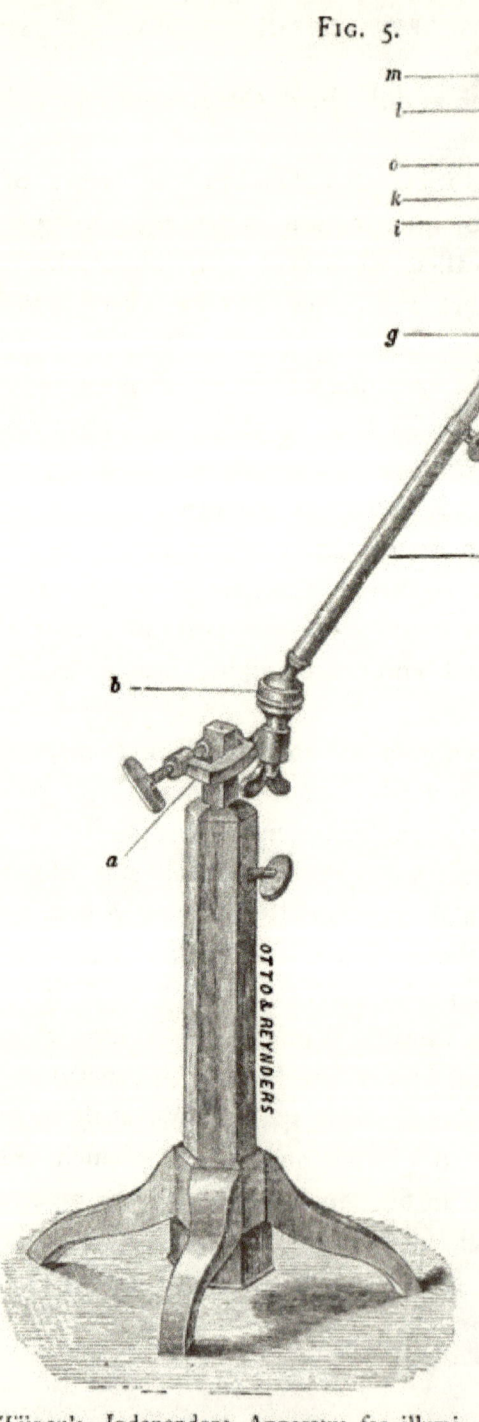

Türck's Independent Apparatus for illumination.

a. A screw to fasten the apparatus to a stand made for the purpose, which may be raised or lowered according to discretion, or to attach the same to the back of a chair.

b. Lower ball and and socket joint.

c. d. Brass tubes of which *d* is enclosed in *c* and can be lengthened or shortened according to desire, and be made fast through screw *e*.

f. Prisma confined within tube *d*, movable.

g. Screw to the prisma.

h. Another flat joint with screw.

i. Cylinder for the attachment of a magnifying glass.

k. Upper ball and socket joint.

l. Reflector.

m. Knob to attach the mirror to the spring supporter.

n. Receptacle with a screw for the reception of the nail-like attachment (*o*) extending from the reflector.

The same apparatus is also manufactured with three brass tubes fitting into one another, instead of only two, as in the accompanying drawing. This is called the large apparatus; both from the fact that the brass tubes fit into one another, are very compact, and the reflector being separated, the whole is easily packed and carried about.

MODUS OPERANDI.—*The Lamp and its position.*

Any lamp, from a study to a common house lamp, answers the purpose, provided the flame be white, which can be regulated through a glass-chimney of proper dimension and sufficient draught.*

The lamp is to be placed behind, and on the right of the patient's head, who is seated on a chair. Its flame ought to be about an equal distance behind the ear, as the latter is distant from the angle of the mouth. This may serve as a rule when the concave mirror—the reflector used,—is 6½ inches in diameter. In proportion as the diameter of the reflector is increased the lamp must be moved further back.

The apparatus itself is in front of the patient and to his right. In the absence of a special stand it can be screwed to the back of a chair, upon which a person is seated to steady the same. The operator seated in front of the patient, adjusts now with his left hand the brass tubes to the required height,—the prisma remaining undisturbed, turns the concave mirror by careful rotations towards the lamp and directs the concentrated

---

* Cusco in Paris, 1861, increased intensity of light, by incorporating Oxygen gas with the flame of a common lamp.

rays of light into the cavity of the mouth to illumine the palate and uvula.

A few well directed motions of the reflector, executed immediately *after* the laryngeal mirror is introduced, will afford a perfect view of the parts to be examined.

It is but just to state, in this connection, that to acquire the dexterity to examine patients promptly and accurately with this apparatus, demands more than ordinary application and tact, as compared with the inventions of other observers ; but once the difficulty overcome, and the left hand practised in the rotation of the reflector, the least motion of which will also result in a change of the picture presented to the eye of the observer—the patient student will be amply rewarded by the result of his investigations, as the tourist who has been climbing a steep mountain-path, after many privations and exposures, is rewarded by the sight of a majestic and unsurpassingly grand landscape.

Dr. Tobold's apparatus, remaining yet to be described, has proved itself in our daily practice as the most easily managed, simple in its construction, compact in form, and answering all requirements. It can safely be recommended to beginners in the art, and cannot be dispensed with when once used. For this, as well as much other valuable labor in laryngoscopy, we are indebted to Tobold.

His apparatus, a representation of which we give below, consists of three convex lenses, *c*, *d*, *g*, (vide figure 6), enclosed in a brass tube *a*, which is screwed to a common study-lamp, and at *b* has a movable arm supplied with a screw, whereby the lens can be brought in close proximity with the flame of the lamp, thereby increasing the intensity of light. The lenses *c* and *d* are in close proximity to one another, and can, for the purpose of cleaning, be removed at *f*, as also can lens *g*, by unscrewing the tube at *h*; *m*, is a brass arm with three joints fastened to the lamp, to the extremity (*s*) of which the handle attached to the reflector (*i*) is screwed; at *o* is a simple Charnier joint, by which to move the mirror back and forward.

It is unnecessary to move the reflector itself, as any side motion of it is easily gained by the movable arm, which acts with perfect facility. A brass rod with screw, an Argand gas-burner, and sufficient rubber tubing, enables us to dispense with the lamp altogether, and to set up and employ this apparatus anywhere where gas has been introduced. Such is the apparatus (besides that of Prof. Türck) we daily employ, and which we recommend.

FIG. 6.

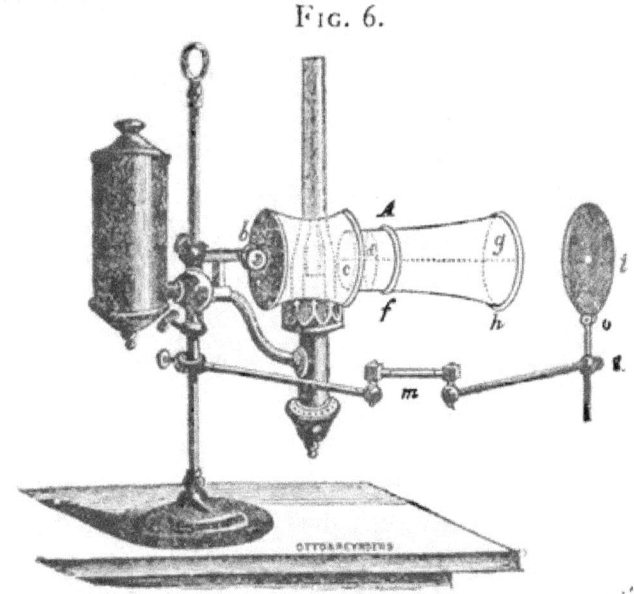

Tobold's Apparatus for Illumination.

Section II.

DIRECT LARYNGOSCOPIC EXAMINATIONS.

Having already considered laryngoscopic examinations by direct sunlight, both with and without reflection, let the room be sufficiently darkened, by drawing down the curtains, and the examination be made by artificial light, for which purpose we shall employ Tobold's instrument.

The following progressive steps will claim our attention.

1st. *Position of patient, physician, and arrangement of the apparatus.*—When artificial light is employed, it is stationed to the right, and somewhat in front of the patient; the centre of the flame being in direct line with the mouth, in order that the rays of light from the reflector may fall in a horizontal line directly over the back of the tongue, upon the laryngeal mirror, held against the palate by the operator. In order, moreover, that the light may fall exactly into the median line, the face of the patient must be turned slightly to the right. The examining physician, with his legs in proximity and outside those of his patient, sits close in front, his eyes in a line of equal height with the upper lips of the former. Then approaching the border of the reflector, he looks into the throat of the patient; when, in order to inspect more critically particular positions of the cavity of the larynx, he bends his head to the side, and makes the examination through the central perforation in the reflector. Intensity of light is thereby gained.

Some laryngoscopists use the centrum seldom, others constantly and exclusively. It is a matter of practice and habit only.

2d. *Position of the head and tongue of the patient.*—The position of the head is important. Bent somewhat backwards, the lower border of the upper lip ought to be in a line with the insertion of the velum palati. Most patients will readily hold the head as directed when it is explained to them, and when they are assured that they are not going to be hurt. Nevertheless, at the first attempt (sometimes oftener), to introduce the laryngeal mirror into the mouth, they involuntarily recede, sometimes quite violently, and a second trial has to be made.

To remedy this evil, a head-rest, an arrangement like that attached to photographers' chairs, has been adopted by some; an admirable contrivance when operations are to be performed within the throat. Except for operations and exceptional cases it will be found, that the physician's left hand supporting tongue and chin, and a little patience on both sides, will suffice to accomplish the object.

The head in proper position, let the patient open the mouth and protrude the tongue as far as possible; then take a towel or soft handkerchief, held in readiness for the purpose, dry the protruded tongue, lay the cloth on it, place the index-finger of the left hand over and across it, and the thumb under it, hold and draw it firmly yet gently over the underlip so as to depress the root of it, whilst the mirror is introduced into the throat. If the patient is fretful and moves the head, let the operator hold the tongue as described; otherwise it is best to let the former go through this manœuvre, as it gives confidence, occupies the mind, and last, but not least, leaves the physician's left hand free. We very rarely assist the patient, and then only at the first examination. A few individuals possess sufficient control over the tongue to hold it down by a voluntary effort, while the laryngoscopic examination is made. Sometimes this power is acquired after considerable practice.

A metallic tongue depressor may be used where the tongue shows great resistance; it may be depressed with one or two fingers of the operator's left hand. We have, however, invariably found that any attempt to depress the tongue is usually less successful than its gentle and steady traction forward, for the reason that when it is depressed in front with force, it is also at the same time pushed backward at the base, and upwards too, touching nearly the back of the pharynx. Sometimes it arches upwards so as to touch the roof of the mouth. The result is, that the passage of light into the larynx is obstructed, brings the tongue in contact with the laryngeal mirror, and thus excites nausea. To depress the tongue with finger or instrument should therefore be avoided.

3d. *Manner of holding, introduction, position, and changes of the laryngeal mirror.**—When the patient and

---

* The laryngeal mirror has already been described at length.

the operator have taken the requisite positions, the apparatus for illumination properly adjusted, the patient's head placed rightly, the tongue protruded and fixed, the physician takes hold of the mirror, which, before being introduced, is first to be warmed by holding it over a lamp or by dipping it into warm water (of 50° or 60° R., for, if the water is too hot, the silver coating of the glass mirror is easily spoiled, and the mirror rendered useless*), so that the patient's breath may not dim it. Its temperature should, however, every time be carefully tested by the operator, by bringing it in contact with the cheek or the hand, for the reasons, that an over-heated mirror will burn the patient's mouth and spoil the silvering.

The mirror is to be held like a pen (see figure) resting between the thumb and side of the first phalanx of the middle finger, so that by simple pressure of the thumb upon the handle, the mirror can be rotated as little or as much as desired. The index-finger does not come into requisition.

---

* The use of warm water has also another advantage, viz., if the mirror is bespattered by sputa, blood, etc., during examination, it can at once be washed, cleansed and warmed again in one and the same process, instead of washing it first in cold water, drying it, and then lastly warming it again over a lamp.

Fig. 7.

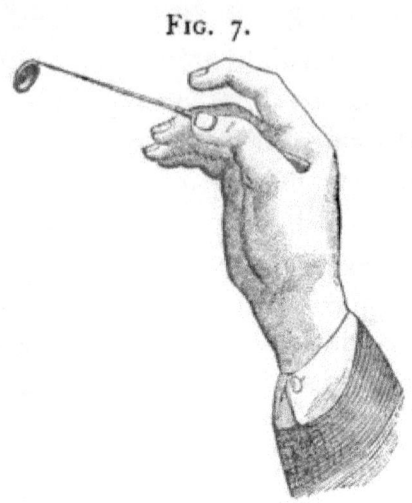

Position of the Hand and Mirror.

The hand should be held naturally, easy, not stiffly bent backward. This requires patient practice. Held thus, it is introduced so as to slightly raise the uvula and soft palate. Take care, lest the uvula project below the mirror, and its image being reflected in the glass it thus obscures the view of the larynx. Avoid touching the tongue, and particularly the back of the pharynx with the mirror, these being the most sensitive parts within the mouth.

In exceptional cases the pharynx bears the touch of the mirror as well as the uvula and soft palate. By resting the third and fourth finger upon the chin of the patient, the hand of the operator may be kept steady, and assist him much to carry his mirror undisturbed over the tongue to its destination. At the same time, the stem of the mirror must be pressed into the angle of the mouth, whereby the mirror, held somewhat obliquely, passes exactly along the median line, and receives additional support.* Having avoided the back of the tongue and the pharynx, and placed the mirror in an oblique position below the soft palate, with the uvula at its back, we at once obtain a view of the larynx, at the same time the patient—as instructed beforehand—is to breathe gently, short and regularly, as if he were moderately out of breath. Practice enables us readily to make such changes in the position of the mirror (by slight rotation executed by the thumb), or of the patient, or in the direction of the light, as may be required, to bring the parts fully in view.

It must be remembered, that the picture of the larynx as it appears in the mirror, is reversed, so that we get the same view as we do when examining the larynx after death. We look at it from behind. The epiglottis is seen first (it shall engage our attention by and by); the arytænoid cartilages appear nearest to the eye; the insertion of the vocal cords into the thyroid cartilage is more distant. The anterior wall of the trachea is seen

---

* For the benefit of beginners, to train the hand and eye, Dr. Tobold has introduced a phantom in the shape of a prepared skull fastened to a portable stand with slide and screw, into which is introduced a plaster of Paris cast of the larynx and trachea, with the tongue protruding, truthfully painted, and admirably adapted for the first practice of the art. A second phantom represents the head with open mouth and protruding tongue, into the neck of which is also introduced a larynx, for the purpose of studying the same inside, and gaining the requisite dexterity with the mirror before experimenting on individuals. In our demonstrations before pupils these phantoms are constantly used, and they are found of great advantage. The use of the above-mentioned phantoms can, however, be dispensed with by procuring a human tongue and larynx with the upper part of the œsophagus, and placing or arranging them in a skull with the lower jaw attached. This can readily be done upon a table, the skull being supported on a few books, or if the above is wanting the head and neck of a sheep will answer the same purpose.

as if we were looking into the tube from behind. To the uninitiated, this reversed picture is, of course, at first troublesome, as everything, so to speak, has changed position. This delusion will soon, however, cease to annoy him, if he constantly remembers, that the parts are on the same side of the observer on which the image appears. What, therefore, is copied to the left of the observer in the mirror picture, is in reality also to his left. Likewise, as the right eye of the patient so also the right half of the larynx is to the left of the observer, and so also appears the right half of the larynx in the mirror to the left of the observer.*

Figure 8, taken from Prof. Türcks' Clinic, represents the larynx as the parts are seen in their natural position; figure 9 represents the parts as represented in the laryngeal mirror during examination.

---

* Türck's Clinic, 1866. The Author has availed himself principally of the illustrations of Prof. Türck's late work for two reasons, first: that he considers them superior to any others published; and secondly, in the hope of directing the attention of the profession to the able works of the distinguished Professor.

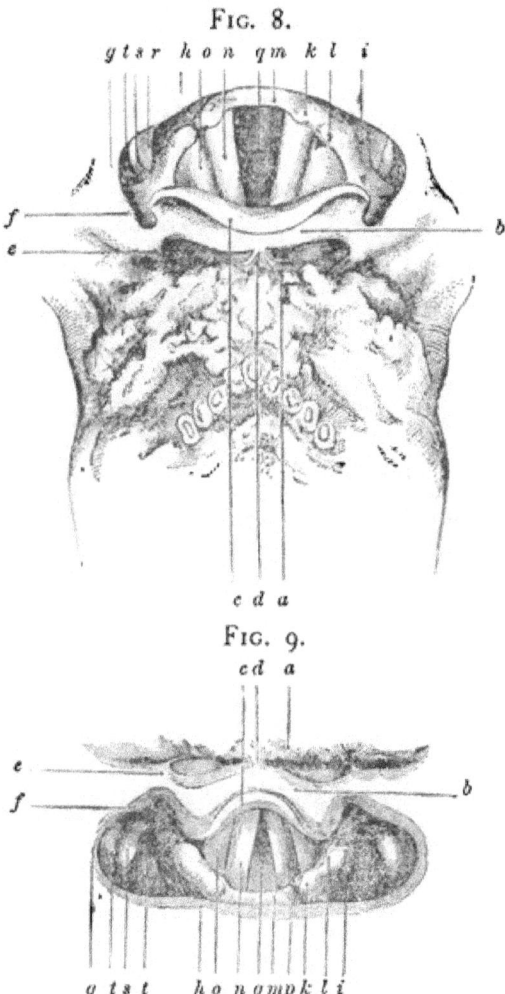

FIG. 8.

FIG. 9.

Fig. 8 & 9. (*a*) Root of the tongue; (*b*) Anterior surface of the epiglottis; (*c*) Its everted border; (*d*) ligam-glossæpiglottic med. with the well known valleculæ (Tourtual) on both sides; (*e*) Right latteral glossæpiglottic ligament; (*f*) Right large cornu of the hyoid bone; (*g*) Right wall of the pharynx; (*h*) Posterior wall of the pharynx; (*i*) Left arytenoid cartilage; (*k*) The cartilage of Santorini; (*l*) Cartilage of Wrisberg; (*m*) Upper extremity of the posterior wall of the larynx, (musc-transversi); (*n*) Right true; (*o*) Right false vocal cord; (*p*) Opening of the left ventricle of Morgagni; (*q*) Glottis, at its anterior extremity the anterior wall of the larynx; (*r*) Right thyroid cartilage, which with the folds of mucous membrane above (*s t*) represents the outer wall of the sinus pyriformis, extending into the outer wall of the pharynx the inner wall of which is formed by the arytenoid cartilage and the ligament aryteno-epiglotticum.

To simplify description, Dr. Gibb (on the Throat and Windpipe, page 452, sec. editn.,) gives the following order of the parts as seen when looking into the throat with the mirror:

1. The back of the tongue.
2. The valleculæ, or fossæ at its base.
3. The epiglottis.
4. Posterior part of the cricoid cartilage, with its mucous membrane.
5. Larynx.
6. The arytenoid cartilages, with their apices, the cartilages of Santorini.
7. The aryteno-epiglottic folds, or ligaments, with the cartilages of Wrisberg in the negro.
8. Vestibule of the glottis.
9. Superior thyro-arytenoid ligaments or false vocal cords.
10. Ventricles of Morgani.
11. The true vocal cords, or glottis.

Beyond the trachea sometimes the bifurcation of the same is distinctly seen.

When the patient makes a deep inspiration, the glottis, which during common or regular respiration is only partially open, is then a wide triangular opening of considerable size, and the vocal cords appear on each side of a pearly white color. If again the patient is requested to pronounce the German diphthong "æ," the glottis closes, whilst the vocal cords approach one another closely, and vibrate with the impulse of the expired air.

Often experiments can be carried very far in this

direction. We have succeeded with patients, particularly singers, to such an extent—after a positive tolerance has been established—as to make them sing (of course without pronouncing the words), short melodies, during the progress of which the opening and closing of the glottis, the variable tension of the cords, and the action which the remaining component parts of the larynx took therein was most beautifully exhibited.

It is equally important to practice the introduction of the laryngeal mirror with the *left* hand as well as with the right. The operator, holding the mirror in his left hand to gain a view of the larynx,—the patient manipulating his own tongue,—can use his right hand for the introduction of brush or other instrument in situ, and thus make his applications with discrimination.

4th. *Obstacles encountered in laryngoscopic examination.*—It remains yet to speak of the *obstacles encountered and to be overcome* in the introduction of the laryngeal mirror, and during its necessary retention, in the throat.

It is rather singular that those unpleasant sensations, cough, retching, and dyspnœa, which we should expect to follow the instant introduction and contact of the mirror, are very seldom encountered. Among the very large number of cases that we have examined for several years we remember but nine—four of retching and five of cough—that obliged us to give up every further attempt for the time. Other laryngoscopists speak equally of the rare occurence of these symptoms. Others more formidable, we meet often, and have to devise means to overcome them :

(*a.*) *The Tongue*, reference to which has already been made, page 128 et seq. Some persons find it impossible to hold the tongue out any length of time. As soon as an attempt is made to introduce the mirror or any other instrument, up it rises to the roof of the mouth and remains there with wonderful pertinacity. Again, some have more than their just proportion of that unruly member, and even when firmly pulled over the under lip, the arch formed is so high as to render it almost impossible to laryngoscope. Such subjects are generally quite nervous too, *patience* becomes, therefore, *the golden rule* with these. The expectant mode of treatment, so popular with some *confrères*, answers well for a couple of days, but then the attempt must be renewed, and repeated till all nervousness is more or less overcome, and the tongue under control. I am in the habit, as also recommended by Dr. Watson, of directing irritable patients to practice by sitting in front of a looking glass with the mouth open, for the purpose of acquiring the habit of controlling the movements of the tongue whilst inspecting it. It is of great assistance at times.

(*b.*) *The Epiglottis*, which, next to the tongue, presents many obstacles to the successful examination of the larynx, varies much in appearance in different individuals. It may be large or small, broad or narrow long or short. Its free border often projects obliquely downwards and backwards, rendering it sometimes impossible to throw the light beneath it. Sometimes its arch is contracted, sometimes one half literally folded over the other half; at other times depressed (approaching each to the centre) resembling a bonnet tied with ribbons under the chin. In fact, the greatest variation exists.

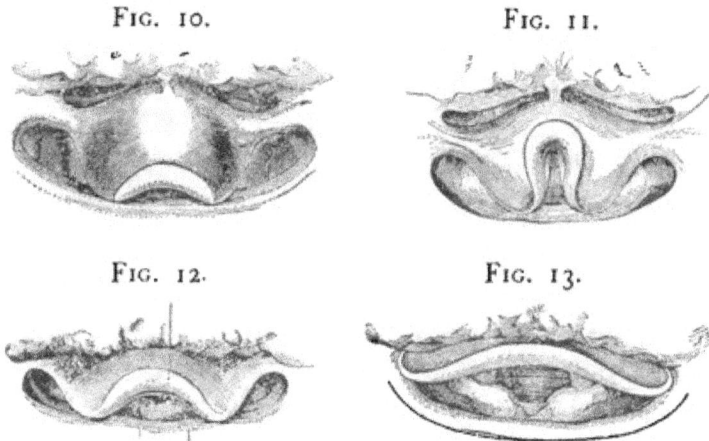

Fig. 10.   Fig. 11.

Fig. 12.   Fig. 13.

Fig. 10 Represents the Epiglottis bent backwards, depressed.

Fig. 11. Omega-shaped Epiglottis, laterally contracted.

Fig 12. The border of the Epiglottis depressed and rolled backwards.

Fig. 13. Epiglottis depressed, but partially raised so as to show the arytenoid cartilages with those of Santorini and Wrisberg, and part of the vocal cords.*

The tubercule of the epiglottis, reaching into the larynx at the anterior junction of the true vocal cords interferes with the sight of that part, even when the epiglottis is in its usual straight position.

The correct position of the head and tongue goes far

---

* Copied from Türck's Clinic.

towards remedying any deviation in the position of the epiglottis. In many cases success is insured when after the laryngeal mirror is introduced, the patient is requested to pronounce the diphthong "æ" several times in succession, each time striking a higher pitch, for instance, the first, third, fifth and eighth note of the scale, when at each successive change of the pitch of the voice, the epiglottis rises step by step, is straightened often at the same time at the base, so that, at a glance, the entire surface of the glottis from the anterior angle to the base can be inspected. With other patients, pronouncing the letter "e" secures the object, whilst again in another class we have to resort to repeated forcible and abrupt inspirations. For exceptional cases means have been devised by Von Brüns and others, to raise the epiglottis, by advocating the use of the epiglottic pincette, or a covered epiglottic hook, which, though extremely rarely required, can be applied without suffering, and renders good service at times.

Dr. Gibb, of London, presented a paper before the British Association for the Advancement of Science, at a meeting at Cambridge in 1862, and at Newcastle in 1863,* wherein he disproved the hitherto received opinion, that the epiglottis is naturally in the correct or vertical position, but on the contrary demonstrated, that in *eleven per cent.* of mankind it was found to be oblique, very much or semi-pendent, or nearly quite

---

* Archives of Medicine for 1863 and 1864, and various journals and papers.

horizontal in persons apparently healthy. His arguments were founded upon an examination, with the laryngoscope, of six hundred and eighty healthy persons, up to September 1863.

By the last census the population of Great Britain was determined to be 28,887,519; eleven per cent. gives the number of 3,177,627 persons who have not a vertical or erect epiglottis. "Can it be wondered at," the author asks, "that diseases of the throat are very prevalent among mankind, when the key-stone to the respiratory arch is shifted in its position?"

It is probable in some countries, those within the tropics for example, the percentage might be even more than has been found to prevail in England."* It would be both useful and interesting, if we possessed some statistics in the United States, in reference to the above fact, particularly in regard to the New England States, where pulmonary and throat diseases are very prevalent

Semelder† states, that in about 25 per cent. of adults he got a perfect view of the larynx easily at the first examination; in about 5 per cent. it was impossible to see the larynx at all; in the remainder he succeeded more or less completely after repeated examinations. This per centage is too large, as, according to Gibb, the epiglottis deviates from its true position in only 11 per cent., which must interfere with the examination. Add to this 5 per cent. more for other causes, and we then have only a total of 16 per cent.

---

* Gibb. Throat and Windpipe, pages 53 and 54.

† Dr. S. Semelder. Vienna, 1863.

It is safe to assert, that in three-quarters of all the patients I have examined, and they number hundreds, I have been able to obtain a good view of the larynx. This estimate does not include children below six years of age.

(*c.*) *Great sensitiveness and excitability of the fauces*, at the touch of the mirror during examination is a further obstacle, as it excites contraction of the pharynx and retching. In some patients this is natural and unavoidable, in the majority however it is owing to a congested state of the mucous membrane of the fauces. Hence, when a throat appears at first sight red and engorged, difficulty may be anticipated in the examination of the larynx. Special care must be taken by the operator to avoid any unnecessary increase of that already troublesome excitability.

After eating or drinking, this is always increased, therefore irritable persons ought to be examined either fasting, or a few hours after breakfast. Again, an increase of sensitiveness is noticeable during the continuance of catarrhal affections of the upper portions of the respiratory and digestive organs, in the same individual in different degrees at different times of the day, according to the general feeling, especially in nervous individuals, so that, when the examination has been made successfully one day, it has to be abandoned the following, for the above reasons. Various modes have been suggested for lessening the sensibility of the throat in such cases. One is, to direct the patient to take a piece of *ice* into the mouth and let it dissolve before examination. This works well.

Bromide of Potassium, either swallowed or used as a gargle—a solution of *Morphine* in *chloroform, hydrochloric æther* and *tannin* have all been tried. Of these the last mentioned, in a solution of 16, 20, 30 grains to the ounce of distilled water, and administered through a pulverisateur (such as Lewin's), has rendered the best service, whilst the bromide has disappointed expectations. Morphine with chloroform is exceedingly painful when applied to irritable membranes; and lastly, hydrochloric æther in diminishing the great sensitiveness causes burning and produces an irresistible desire to cough. A more successful way is to put 10 to 20 drops of chloroform on a handkerchief, and let the patient inhale it for a minute. I find it quiets the most irritable throats. The number of drops used is too small to render the patient drowsy. Yet none of these appliances are so effective as the repeated introduction of the faucial mirror at intervals of a day or two.

(*d.*) *Hypertrophy of the Tonsils* may render the examination of the larynx difficult or impossible. In proportion to this enlargement, the mirror must *be* adapted to the examination. When the tonsils are, however, so much enlarged as to touch each other, laryngoscopic examination is impracticable. They ought first to be removed by the guillotine, or by the application of Vienna paste.

(*e.*) *Irregularity of Respiration* must finally be mentioned as one of the causes upon which failure of laryngoscopic examination sometimes depends. Most patients think, that after the head and tongue are fixed, it is also necessary for them to retain as long as possible their breath, instead of breathing quietly and regularly without exertion. Nothing but a careful explanation of the respective steps to be gone through with, on the part of the operator, can insure success; whilst at the same time it will be found that patients laboring under serious diseases—such as are attended with suffering—bear the examination better than those who have but trifling ailments or none at all.

CHAPTER III.

Auto-Laryngoscopy.

To enable us to judge of the sensations of others when examined, to acquire skill and confidence in our manipulations of the larynx, and for the purpose of gratifying our patients and our curiosity, *Auto-Laryngoscopy* ought to be practised by every one who wishes to be proficient in this art. The methods of examination are various.

Garcia's, the father of auto-laryngoscopy, is very simple. Turning his back toward the sun, he received the solar rays by means of a small mirror held before the face, and directed them upon the laryngeal mirror placed against the uvula, which reflected again the image into the other mirror.

Moura Bourouillion's laryngoscope is available for these examinations, though little used.

Dr. G. Johnson proposed the following simple and satisfactory plan: "Sitting at a table of a convenient "height," he says, "I can place a looking-glass at a distance of about eighteen inches in front of me, and a moderator lamp on one side of the glass, but two or three inches further back, so that the light may not pass directly from the lamp to the mirror. Now, with the reflector on my forehead, I direct the light as it were into the open mouth of my own image in the looking-glass; then, introducing the laryngeal mirror into my mouth, I see the reflection of my larynx and trachea in the glass before me, and any one looking over my head can see the image at the same time. By this method the experimenter can see his own larynx and show it to others." We have tried the above method and found it as successful as it is simple, so that it can be carried out with facility by beginners. Omitting to speak of several other apparatuses for Auto-laryngoscopy, we shall only refer to Prof. Czermak's, as by its use in his own hands the art has become, so to speak, naturalized in every foreign land. It consists of a mahogany box, at the bottom of which is a sliding panel, to which is screwed a brass tube, which permits of the attachment of a large illuminating concave perforated reflector, at any height most suitable. Opposite to this is a receptacle for another tube, which latter receives the stem of an oblong mirror, attached by means of a hinge. The light is placed to one side of the experimenter, and throws its rays into the large mirror, which reflects him in the laryngeal mirror held against the uvula in the right hand. The observer looks through the perforation, or around the margin of the large mirror, whilst the experimenter looks into the oblong mirror; both see the laryngoscopic image in the laryngeal mirror, although not precisely alike to each, as their visual axes do not form the same angles with the reflecting surface of the mirror. From the position of the lamp the eyes are completely protected from the glare.*

This apparatus has this advantage, that a large party of persons can see the laryngoscopic image by clustering around the reflecting mirror, and at the same time others, by looking into the oblong mirror, will see nearly the same object. For general convenience, handiness of arrangement, regulation of light, elegance, and simplicity, this apparatus is to be recommended as preferable to any other, at a cost which places it within reach of all.

For lecturers on physiology, Czermak's Auto-laryngoscope is most indispensable, and should have the preference over every other instrument. Such is the language of commendation of Dr. Gibb, one most able to judge in such matters.

---

* Gibb on the Throat and Windpipe, p. 458 et seq., Sect. Auto-Laryngoscopy. *Medical Times and Gazette*, Feb. 14, 1863, p. 157.

CHAPTER IV.

RECIPRO-LARYNGOSCOPY.

THIS, the art of demonstrating a patient's larynx to others, was first pointed out by Dr. Smyly, of Dublin.* He has contrived to overcome the difficulty of showing the larynx to a third person in the ordinary way. To quote his own words: "In the ordinary method, when the examiner has a full view of the vocal cords of the examinee, he calls upon his colleague to view the parts, who when he places his head beside that of the examiner, only gets a partial view—a portion of the epiglottis, one arytenoid, and perhaps a vocal cord. In endeavoring to see more he pushes the examiner's head, so as to displace the light, or shakes his hand, so as to bring on nausea. Many other inconveniences will occur to the mind of the practical laryngoscopist, which I shall not here allude to.

" My addition consists of a simple square piece of very good plate-glass mirror, set in brass, like the ordinary concave mirror. A second split tube is soldered on close to the tube, which exists on all Weiss' fron-

---

* Dublin Quarterl. Journal, Vol. XXXVI., Aug. 1863.

tal bands,* and a brass rod, the ends of which are bent in opposite directions, at an angle of 45°.

"The mode of using this glass is as follows: The laryngoscope is fixed, as usual, before either the left or right eye. The brass rod is fixed in the tube, beside that which holds the rod supporting the reflector, and my square glass is fixed on the other end, as is very well shown in the engraving.

Fig. 14.

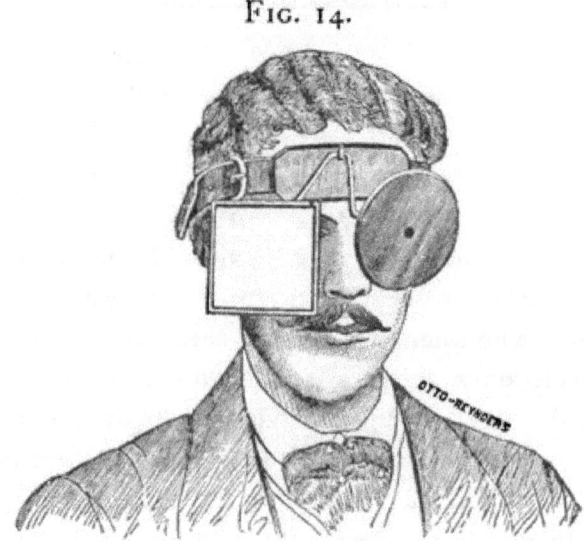

Dr. Smyly's Recipro-Laryngoscope.

"As the angles of incidence and reflection are equal, the mirror may be turned to such an angle that the second examiner may be placed at such a distance from both the patient and operator, that his presence cannot disturb the steadiness of either. The view the second examiner has of the larynx in the square mirror is not

---

* Also to be had at Otto & Reynders', Chatham street, New York.

inverted, being twice reflected. The right vocal cord of the examinee is to the right-hand side of examiner number two.

"The glass employed in the manufacture must be as perfect and parallel as possible, so that the loss of light may be a minimum.

"In conclusion, I may add that the additional weight of the square glass, when made in an artistic manner, is scarcely perceptible."—*Dublin Quarterly Journal*, Vol. XXXVI., Aug. 1863.

CHAPTER V.

INFRA-GLOTTIC LARYNGOSCOPY OR TRACHEOSCOPY.

FIRST suggested by Dr. Neudörfer in 1858, but first carried out in practice by Prof. Czermak in 1859. It consists in the introduction of a very minute mirror through the opening made for the insertion of the canula in tracheotomy. The face of the mirror being directed upward, a view of the larynx is obtained from below. Several cases, one by Semeleder, of this successful mode of operating, have been reported. We had, about a year ago, the rare opportunity of examining a discharged soldier, on whom, two and a half years before, tracheotomy had been performed in consequence of a gunshot wound received in battle, and who wore a canula. Having introduced our little mirror of Neudörfer, by very strong reflected light we could see perfectly the lower surface of the vocal cords, of a reddish color, in contradistinction from the upper surface, which was pearly white, and could be distinctly seen; as the epiglottis was not bound down by cicatrices, as ordinarily happens.

RHINOSCOPY.

CHAPTER VI.

ALTHOUGH Bozzini in 1807, Baumes in 1838, and Avery in 1846, practiced the art of examining the posterior recesses of the nostrils and of the pharyngo-nasal recess, yet it was not until Czermak,* in 1859, in his first publication on the laryngoscope, called attention to the fact that the same method of examination was applicable to the posterior surface of the soft palate, the posterior openings of the nasal fossæ, and the superior parts of the larynx. He also contrived various useful instruments for this purpose. Prof. Türck, Semeleder, Störck, Voltolini, Wagner and others, have since made many valuable contributions to and improvements in this art.

To accomplish what is claimed for Rhinoscopy requires patience and perseverance, and more so than laryngoscopy, as its practice is more difficult. Unsatisfactory results, especially in beginners in the art, are generally, with few exceptions, owing to want of perseverance, sometimes, however, to lack of skill.

We shall describe the instruments necessary and the *modus operandi* during examination.

There are required :

---

* Wiener Med. Wochenschrift, Aug. 6, 1859.

1st. *A small mirror* of glass, smaller than usual for laryngoscopic purposes, about five-eights of an inch in diameter, and placed at a right angle to the shank.

2d. *A reflector*, which may be the same as used in laryngoscopy.

3d. *A tongue-spatula*, which has been made of various lengths and at different angles. Prof. Türck's spatula, recommended especially for rhinoscopic purposes, has proved itself the most practical adopted, from the fact that the stem or handle is separable from the spatula blade itself, so that a larger or smaller-sized blade can be screwed to the handle, and the point of distance of the blade from the handle can be regulated at option. Thus it can be used for children and adults. It is also easily packed and carried about.

[See Figures 15, 16, 17, 18 and 19 on the following page.]

4th. *A palate-hook*, for raising the uvula and pulling it forward, about four inches long, narrow at its fixed end, broader and somewhat curved at the opposite end. Recommended by Czermak. This instrument is now seldom used. Türck advises the use of a loop, devised by himself, to secure and keep the uvula out of the way. Figure 20 is a representation of his palate lasso, or loop.

LARYNGOSCOPY AND RHINOSCOPY.   51

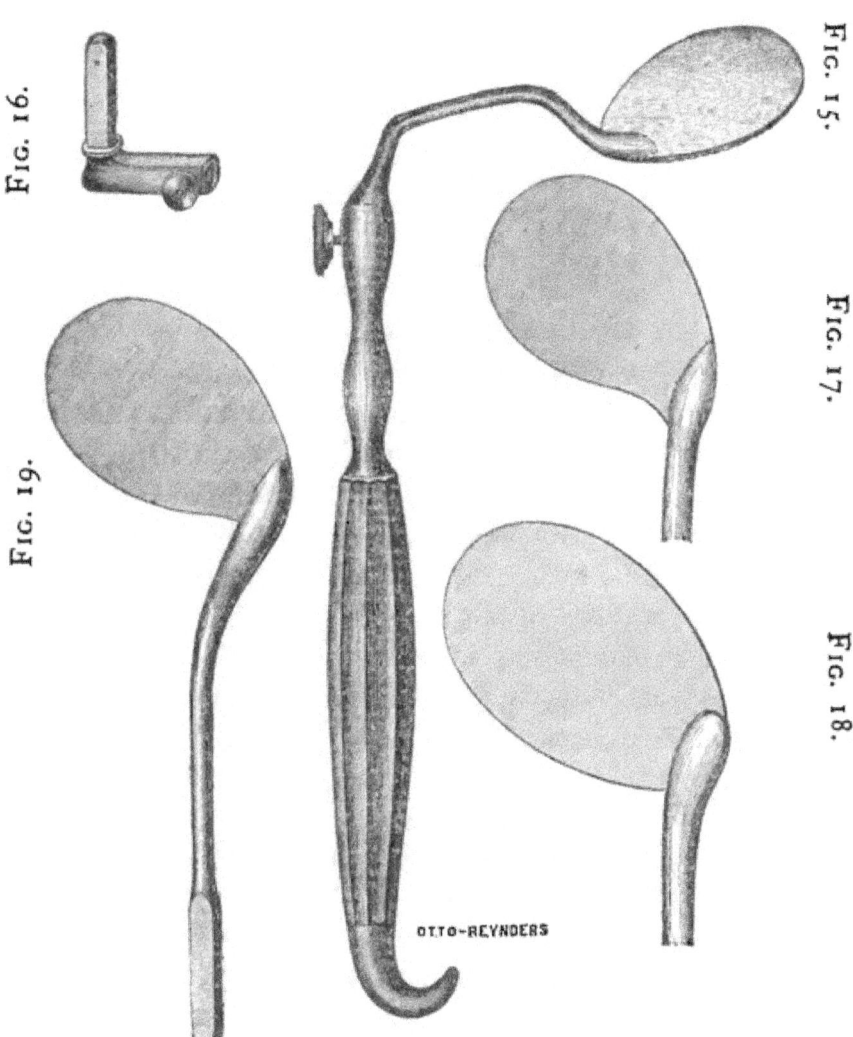

Fig. 16. Fig. 15. Fig. 17. Fig. 18. Fig. 19.

Türck's Tongue-Spatula.

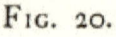

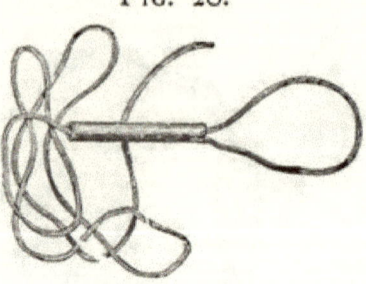

Palate Lasso.

It consists of a small silver tube, about one inch long, and sufficiently wide to pass a double thread through it. Near one end of it is a hole, through which is passed a double well-waxed thread, fastened by its centre. When the other end of the double thread is again passed through the tube, a sling is formed, the object of which is to fasten the uvula. This is accomplished by holding the tube with the loop in a pair of forceps, then bringing the loop under and on to the uvula. It is fastened by pressing the tube firmly toward the uvula, and holding the thread behind stretched whilst it is fastened to a hook attached to an elastic band tied around the patient's forehead. This process, though at first rather troublesome to be accomplished with ease, secures the uvula without pain, and keeps it temporarily out of the operator's way. To remove the loop and tube, it is simply necessary to pull the tube with the forceps backward, when the loop drops from the uvula. Whatever may be said for or against the employment of a hook, or any other device, to secure the uvula, we must acknowledge that we make our most successful examinations without a hook or loop. Sometimes the use of the hook is attended with discomfort; its contact excites contraction of the palate, which is then drawn upward and backward, so as to obstruct the view completely.

It will be seen at a glance, in looking into the patient's mouth, whether the examination can be made without hook, loop, or any other contrivance to keep the uvula in its proper position. This depends upon the space which exists between the palate and the posterior wall of the pharynx. Whenever the interval is a moderately wide one, the mirror can be introduced without touching the uvula or palate.

During the examination the patient must sit erect, without throwing the head back, while the light is thrown into the mouth by the reflector, illuminated by a lamp or gas-burner, placed in the same position as in laryngoscopy. He is then directed to open his mouth widely, whilst the tongue is pressed forward and downward by a metallic spatula, either held by himself or the operator. The patient being directed to breathe quietly, the mirror is then introduced by the side of the uvula, beneath the palate, with its surface directed upward and forward, so that the plane of the reflecting surface forms with the horizon an angle of about 130 degrees. After introducing the mirror, the observer can steady it by resting his third and fourth fingers on the patient's lower jaw. (Dr. Mackenzie and Voltolini of Breslau, both use an instrument in which the tongue-spatula and mirror are combined in one.)

The following objects will then be successively presented to view:

(*a.*) The septum narium;
(*b.*) The posterior orifices of the nasal fossæ;
(*c.*) The middle and lower turbinated bones;
(*d.*) The orifices of the Eustachian tubes;
(*e.*) The roof of the pharynx, and sometimes
(*f.*) The floor of the mouth.

Figures 21 and 22, Türck's work, are reproduced here for the sake of illustration.

FIG. 21.

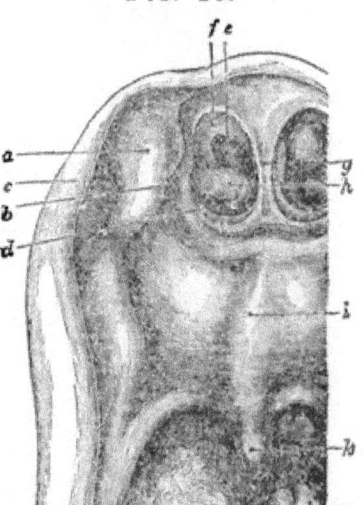

Fig. 21 represents a posterior view, after removal of the dorsal vertebræ of the anterior wall, and a part of the side-walls of the naso-pharyngeal space.

a, Tubercule of the left Eustachian tube; *b*, its opening; *c*, fossa of Rosenmüller; *d*, inferior; *e*, middle meatus; *f*, middle; *g*, inferior turbinated bone; *h*, posterior extremity of the septum between both turbinated bones; *i*, posterior surface of the soft palate; *k*, uvula; *l*, dorsum of the tongue.

Fig. 22.

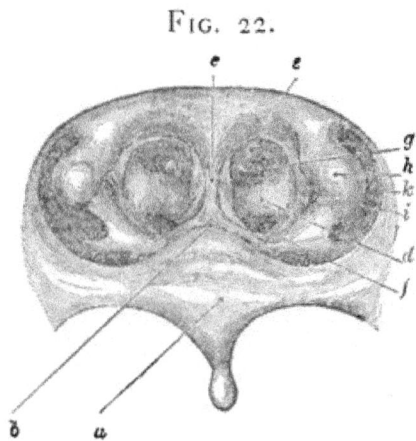

Fig. 22. Rhinoscopic representation of the anterior and a part of the lateral walls of the naso-pharyngeal space; *a*, posterior plane of the soft palate, with the uvula looped up; *b*, point of union of the soft palate; *c*, septum between both turbinated bones; *d*, inferior; *e*, middle left turbinated bone; *f*, inferior; *g*, middle left meatus; *h*, tubercule of the Eustachian tube; *i*, opening of the tube; *k*, fossa of Rosenmüller.

These component parts of the picture presented deserve a most careful examination and ought to be vividly impressed upon the memory to avoid mistakes in diagnosis of disease of those parts. The septum, seen first, is rarely symmetrical, but slants a little to one side, usually the left.

The middle turbinated bones, projecting from the outer wall of the naris on each side toward the septum, are covered with a pale mucous membrane, resembling polypi; so much so, owing to their form and color, they have been mistaken for such until the eye of a more practiced observer settled the matter by means of a rhinoscopic examination. The superior turbinated bones, small and somewhat triangular in shape, are distinctly seen. The inferior turbinated bones are seen as two pale, roundish, solid-looking tumors at the base of the nasal fossæ. Of the three, the middle projects the furthest. Of the three *meati*, the superior one is the largest, whilst the inferior one appears only as a thin dark line.

The orifices of *the Eustachian tubes* are two irregular openings, looking downward and outward, on each side of the turbinated bones, though further back and in a different plane. Buried in the pharyngeal wall, of a lighter color and more yellow tinge than the surrounding mucous membrane, the opening of the Eustachian tube is trumpet-shaped and bevelled off at the upper and posterior edge. Beneath the nasal fossæ is the soft palate, uvula, etc.

As in certain cases it is impossible to make a rhinoscopic examination, which can be easily determined by examining the fauces, so on the other hand this examination is much facilitated in cases of fissure of the soft or hard palate, or loss of the former by ulceration.

Occasionally small mirrors are introduced into the front of the nose, in order to examine the nasal cavity and inferior turbinated bones. It is stated, also, that in persons with capacious nostrils, the nasal orifice of the lachrymal canal can be seen.

Though rather difficult of application and limited in extent, very valuable information may sometimes be obtained by rhinoscopy, in cases of *ozæna, the various forms of ulceration of the hard and soft parts at the back of the nose, in obstruction of the nasal passages by polypi or thickened mucous membrane.* It also enables us in cases of deafness dependent on obstruction of the Eustachian orifice, to diagnose the affection correctly, and to introduce the Eustachian catheter with safety and precision.

We shall embrace the opportunity to introduce here a few cases, to illustrate the practical value of rhinoscopy.

CASE I.—*Obstruction of the right nasal passage from thickening of the mucous membrane of the right middle turbinated bone. The right nostril closed for twenty months. Relief by local treatment.*

Mr. C., æt. 34, merchant, presented himself in January, 1866, on account of a constant stopping up of the right side of his nose, accompanied at times with a very troublesome sensation of dryness, and at times again by a free watery discharge. He felt as if he had constantly a cold in the head, and sniffed incessantly. Took medicine internally, and solutions of zinc, alum, nitrate of silver were sniffed and also injected into the affected parts without benefit.

Condition for rhinoscopic examination favorable.

Result of Inspection.—The mucous membrane of both nostrils was of a red color, of deeper red on the right side, where that part of it covering the right turbinated bone was so much swollen as to nearly block up the entire right nasal passage. I applied solutions of nitrate of silver, sulphate of copper and tannin with a curved brush, for some time, with little result, when I resorted to iodine and glycerine, equal parts, which proved successful in a comparatively short time. The swelling disappeared, and there has been no return of the disease.

CASE 2.—*Ulcers on the vomer, the right, middle and left inferior turbinated bones, resulting in ozæna of three years' standing.*

Mr. F., æt. 36, salesman, had been subject to a very offensive discharge from the nose and throat for nearly three years, which latterly grew so offensive as to be intolerable. Being of strumous habit, the tonsils moderately enlarged, but not so as to obstruct the space between the velum-palati and pharynx, rhinoscopic examination showed the existence of an ulcer near the base of the vomer and inferior turbinated bone on the left side, one on the middle and right turbinated bone. The mucous membrane was of a dirty gray color near the ulcers, for the rest, red.

Treatment.—Solutions of nitrate of silver applied with a sponge fastened to a curved holder, the internal use of iodide of potassium, and regulation of diet. A complete cure in eight weeks.

CASE 3.—*Lodgement of a glass bead in the left nostril of a girl four and a half years old. Removal by forceps.*

A little girl was brought to me, November 1866, who, whilst playing with other children with beads, stuck one into her left nostril. Unable to expel it, she was subjected to a rhinoscopic examination. It was found to have lodged firmly back in the middle meatus. By the aid of the mirror and forceps it was successfully removed.

CASE 4.—Examined Mr. F., æt. 52, May 14, 1867. *His right nostril was obstructed and he had been told that he had a polypus which was the cause of the obstruction, and was advised to have it removed.* Inspected the fauces and found this patient well adapted for rhinoscopic examination. Without any difficulty whatever gained a perfect view of the posterior nasal fossæ and turbinated bones. No polypus was found, but the right middle turbinated bone was thickened and diseased, being the sole cause of the obstruction.

Dr. G. Johnson, in his lectures on the Laryngoscope,* relates a case similar to the foregoing, where the surgeon had already made an unsuccessful attempt to remove a supposed polypus with the forceps, when rhinoscopy proved it to be an affection of the turbinated bone and not a polypus. Another good illustration of the value of rhinoscopy in correcting an erroneous diagnosis is given by Czermak in the last German edition of his monograph† on the subject. A young man had a tumor at the back of his left nostril, which to the touch gave the impression of a polypus, and rendered him deaf on that side. An operation was contemplated, but a rhinoscopic examination discovered a temporary swelling of the mucous membrane nearly as thick as the finger, surrounding the orifice of the Eustachian tube, also great swelling of the middle and inferior turbinated bones, but no polypus, nor any tumor which an operation could have removed or lessened.

The foregoing cases will suffice to illustrate the importance of acquiring the art of rhinoscopy, both as an unfailing guide in diagnosis and in treatment. Be it

---

* London, Hardwicke, 132 Piccadilly, 1864.
† Opus cit., pp. 127-8. Leipzic, 1863.

however remembered, that it requires even more patience in this than to practice laryngoscopy. As Czermak justly remarks: "Many a physician and student has thrown down in despair his mirrors, and quitted the study of this art, on account of a want of patience, so necessary in all great things."

CHAPTER VII.

CONCLUDING REMARKS ON THE PRECEDING CHAPTERS.

1st. It is absolutely necessary, in order to study practical Laryngoscopy and Rhinoscopy with success, that the student should be perfectly familiar with the anatomy of the parts he has so constantly to deal with of their appearance in a healthy condition, before he is able to judge of their diseased state. A description of the larynx, its functions, and the parts intimately connected with it, can be found in every good work on anatomy and physiology. It has therefore been considered unnecessary to introduce the subject here.

Better, however, than the study of plates is it, to familiarize oneself on the dead subject, with the parts in situ; to introduce the mirror, and to dissect the separate parts. Where this cannot be done, a larynx, with tongue and œsophagus attached, will answer the purpose.

2d. The best method, and one not less essential, to familiarize oneself with the shape, position, color and movements of the throat, larynx or nose, and relative connection of all component parts, is to embrace every possible opportonity to examine the throat and larynx of every healthy person, old and young—even children—who can be induced to submit to it, which generally can be done without much persuasion. It is surprising how easy a self-interest can be awakened in unprofessional persons. For over two years I have made it a business to examine the larynx of every person that comes sufficiently long within my presence to enable me to do so. Thus I have examined some days from 2 to 6 persons, including children (exclusive of patients), and noted the condition of the larynx, throat, and particularly the epiglottis, in over three hundred persons. I know of no more beautiful object than the vocal apparatus, particularly that of children. An interest, nay a passion, is developed during these examinations, executed con amore, which will overcome all difficulties.

3d. The color of the parts differs, as described by Gibb.* In a healthy condition, the epiglottis is of a pale salmon or buff color; the interior of the larynx above the glottis is of a pale rose; the true vocal cords are white with a gray shade; the part immediately below the glottis is pale fawn, which in the trachea shades off into a drab; the ring of the cricoid, and the rings of the trachea, appearing of a lighter and almost

---

* Gibb, op. cit.

white color through the transparent membrane. In some persons the arytenoid cartilages are of a yellow pink, and those of Wrisberg, when present, (almost always in negroes), have a yellowish tinge like a small abscess.

At the back of the nose the turbinated bones possess a pale pink and drab color, but when congested, the vessels are generally of a vivid pink, and prominent. The oval trumpet-orifices of the Eustachian tubes generally are pale yellow.

4th. In order to overcome all the difficulties encountered in introducing and placing the laryngeal mirror properly, and to be sure that no part be overlooked during the inspection, or mistaken one for another, let the examination be performed in a strictly systematic order, step by step, as laid down at length in the preceding chapters.

5th. Do not attempt at the first glance to see the vocal cords in the depth of the larynx, but observe first the parts above in succession, and afterward the parts situated lower.

6th. After the mirror is introduced, observe strictly the median line from before backward, the condition of the back of the tongue, the epiglottis, the posterior wall of the pharynx; then let the mirror be turned to the side in order to see the walls of the pharynx, the aryepiglottic folds, and return again to the pharyngeal wall and the epiglottis.

7th. Lastly, is to be examined the interior of the larynx when first its posterior wall is seen; observe the form, size, color, position and movability of the arytenoid cartilages, then lower, the false and true vocal cords in all their relations and power of motion, finally, the posterior plane of the anterior wall of the larynx, from the free border of the epiglottis down into the trachea.

PART SECOND.

CHAPTER I.

APPLICATION OF REMEDIES TO THE LARYNX AIDED BY THE LARYNGOSCOPE.

It has been said without exaggeration, that the laryngoscope has rendered the diagnosis of the diseases of the larynx more simple and certain than the diagnosis of the diseases of any other internal organ; it is equally true as a consequence, that our treatment of these complaints has been placed upon a new basis, the positive basis of a correct diagnosis. When, now-a-days, as Dr. Tobold remarks, a patient presents himself with sore-throat, we grope about no longer in the dark; we are not satisfied with purely subjective appearances, and prescribe some old-fashioned formula for laryngeal catarrh and hoarseness, or even, when these fail, send the patient to a distant water-cure or summer resort, from whence he probably returns little or not at all relieved; but we examine first, mirror in hand, the diseased organ, and apply according to indications our topical remedies, or the knife, where medicines are useless. The treatment of the diseased larynx has therefore been transplanted from the field of medicine to that of surgery.*

Nor is the result of the treatment less satisfactory in this class of diseases since the introduction of laryngoscopy, with the exception, perhaps, of tuberculosis; but even there laryngo-therapeutics afford, sometimes at least, temporary relief.

The various degrees of inflammation of the laryngeal mucous membrane, of the submucous tissues, or the muscles in hyperæmia, affections of the perichondrium and of the cartilages, swelling, and extension of the false vocal cords and aryepiglottic folds, infiltration, ulceration, new growths accompanied by disturbed functions or alteration of voice, all these diseased conditions are amenable to local treatment. Although we will not deny that internal remedies may now and then act as valuable adjuvants, yet it can no longer be questioned, *that the local treatment is the principal factor therein*, since good results are arrived at in much less time, and a radical cure is brought about only through their means.†

It is presumed, that the introduction of the mirror with the left hand has been practiced, so as to be executed with the same care, ease, dexterity and certainty as with the right hand, which now must be employed for the brush, or porte-caustic, or any other instrument called into service. Here again our motion must be quick and certain.

To avoid any illusion regarding the position of the

---

* Tobold, op. cit., page 57.   † Tobold, op. cit.

parts, their respective distance from each other must be carefully measured. Let the patient be reassured, and his head so fixed as to contribute to the success of the operation.

The remedies applied for the cure of diseases of the larynx by aid of the laryngoscope may be divided into three classes :—

1st. We can produce by the direct application of remedies to the diseased spot or part alterative, astringent, or sedative effects, or what is more common, we can cauterize the part, or destroy portions of it altogether.

2d. By the use of instruments, we can bring about a mechanical separation, a destruction, or an entire removal of the parts affected.

3d. Galvanism may be applied directly to the mucous membrane of the larynx, for the twofold purpose of stimulating muscular contraction, and for destroying a certain grown part by the electric cautery.

Each of the above three classes shall be considered seriatim.

CHAPTER II.

REMEDIES directly applied to the larynx may be in the form of:

    a. Solutions.
    b. Powders.
    c. Solid caustic.
    d. Escharotics.

SECTION I.—SOLUTIONS.

When applied:

1st. In more or less diffused hyperæmia of the mucous membrane and submucous cellular tissues, especially in chronic catarrhal conditions which do not yield to milder remedies in the form of inhalations;

2d. In pseudo-membraneous formations upon the free mucous membrane, the well-known exudations in croup and diphtheria;

3d. In hypertrophy of the mucous membrane, involving its whole structure, or only its superficial layer and the epithelial cells, as has been observed along the free border of the vocal cords;

4th. In superficial abrasions and ulcerations which may extend over more or less surface.

In the diseases included in the foregoing four classes, the question is not one of a *deep-seated abnormal* condition, but one of greater or less extent, generally confined to the surface, which through the application of stimulating and alterative remedies is to be brought back to its normal condition. (Von Brün's Laryngoscopic Surgery.)

Remedies used.—Among the most efficacious of these must be classed *nitrate of silver*, which, from its general utility and great importance, deserves more than a passing notice.

The solutions used vary from 1, 2, 4, 5 scruples to the ounce of distilled water. A two-scruple solution will be found the most serviceable on all ordinary occasions, for it is unquestionably agreed among all trustworthy observers, that for the purpose of application into the larynx, trachea, fauces, or nose, a solution of less strength is only trifling with the patient and produces no satisfactory results. For the treatment of ulcerations of the epiglottis or about the larynx, solutions of greater strength are fully warranted. Whilst, however, some are timid, and for fear of doing injury would confine themselves to five or ten-grain solutions, others are not wanting who advocate the use of the solid nitrate, and the strongest concentrated solutions to an ulcerated or otherwise sore throat. These latter must be warned of the risk they incur of destroying the tissues by means of their remedies. The larynx is too delicate a structure and of too great importance to its possessor, to admit of its being placed in jeopardy by unsafe and destructive remedies.

Other valuable remedies are tannin, perchloride of iron, sulphate of zinc and copper, iodine, alum, bromide of ammonium, hydrochlorate of morphia, carbolic acid. The most useful solvent for these agencies is glycerine, by reason of its being viscid ; it adheres to the surface of the mucous membrane, and retains there the remedy it holds in solution. But this solvent property of glycerine is subject to certain proportions. Tannic acid dissolves in glycerine only in proportion of two drachms to the ounce. This forms one of the most useful topical astringents ; with nitrate of silver, sulphate of zinc and alum, it will combine by aid of heat in the proportion of two drachms to the ounce. It dissolves also one-fifth of its weight of hydrochlorate of morphia, a weaker solution of which is a very useful application in irritation of the larynx. Which of the above remedies to apply in each case is a matter of judgment.

How applied.—Solutions are applied by means of the *laryngeal brush, the sponge-carrier, the syringe or injector,* and *pulverisateurs for inhalation.*

1st. *The laryngeal brush,* which has now mostly superseded the sponge recommended by Watts, Greene, and others, was first proposed in its present and most useful form by Professor Türck. A large, full-bellied squirrel's or camel's hair brush, cut square at the end, is firmly attached to a silver stem from four to five inches long, of sufficient strength, so that said stem can be bent at any convenient angle between $90°$ and $102°$, according as we wish to touch the anterior insertion of the vocal cords, or the arytenoid cartilages, or any other part of the throat, as circumstances may demand. The stem is fastened to a handle of a convenient shape and size,

70   THE PRINCIPLES AND PRACTICE OF

similar to that of the laryngeal mirror. To this brush we confine ourselves mostly, satisfied that the sponge often produces irritation, if not injury, when it comes in contact with the delicate membrane of the larynx.

Fig. 23 represents Türck's laryngeal brush. The brush is firmly attached to the stem, but I have had the same made with a screw at the extremity of the stem, so as to change the brush whenever desired.

Fig. 23.

Dr. Gibb recommends a brush attached to a bent whalebone, as represented in the annexed Fig. 24.

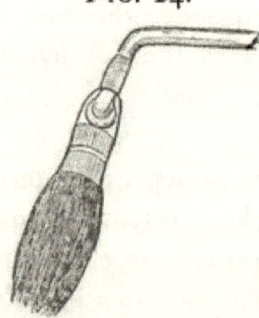

Fig. 24.

Türck's Laryngeal Brush.   Gibb's Laryngeal Brush.

Both Türck's and Gibb's brush can be so constructed that it can be attached by means of a screw to the stem, and thus changed for each patient, and retained for future use, whilst the stem and holder can be equally used for all.

An excellent instrument as Gibb's brush is, yet for certain purposes, I find it objectionable, because the angle of the whalebone to which the brush is attached, cannot be changed at will (only by being seated), when occasion requires it. It is true, this objection may be obviated by having on hand brushes inclined at different angles and also of different sizes.

It will hardly be necessary to suggest that each patient be provided with a separate brush, which after being used is marked with the name of the owner and laid aside for future use on the same person only. After using a brush it should be immediately immersed in water—near at hand—in a glass or bowl, in order to free it at once from any mucous or other impurities which may attach to it; and thus it is already prepared for future use.

The brush is introduced during active respiration, or when the epiglottis is depressed and in order to raise it, whilst the patient pronounces the sound "*ae.*"

When we have passed over the root of the tongue, the brush is then placed against the posterior plane of the epiglottis and pressed almost vertically deeper, in order to reach the glottis. This operation is not so readily performed as it seems, particularly not so quickly as requisite; it requires, therefore, some practice. It is much easier to get the brush behind the epiglottis and press that down, instead of getting over it into the larynx.

When the brush has reached the glottis, violent spasms, fits of coughing, and shortness of breath, sometimes seize the patient and frighten him; but these sensations are readily suppressed by taking a drink of cold water immediately after. To illustrate the result of this simple treatment in acute and chronic inflammation of the larynx, we shall insert a few instructive cases selected from among many. If more evidence is wanted, medical journals contain sufficient proof of cases treated successfully by topical remedies.

Be it owing to our climate—the sudden change of temperature, our mode of heating our houses, our style of dress, our habits of living—certain it is, that the various catarrhal affections are most frequently met with in daily practice. Nor ought it to create surprise, that the larynx, supplied with nerves and endowed with exquisite sensibility, the door-keeper of the lungs as well as the guardian of the vocal apparatus, with its many intimate connections, the pharynx and nasal centres above, the trachea and bronchi below, should be subject to complications, most of which have remained hidden from our view, and thus rendered treatment almost empirical.

Case I.—Rev. J. H.——, D.D., L——, Kentucky, æt. 45, short, active, robust, and enjoying good health; whilst laboring assiduously in the cause of the Sanitary Commission in the West during 1862, '63, '64, was attacked in the spring of 1864, with pain in the fauces and larynx, at times sharply, sometimes dull and heavy, so that gradually he was obliged to leave the parish, and concluded to visit Europe for the purpose of finding medical aid. In Paris, he consulted Dr. Fauvel, who made local applications to the part, gave inhalations, and also sent him to the springs. In November, 1865, the patient returned benefited, but still complaining of the pain, and inability to speak for any length of time, he placed himself under my care in the city of New York.

Laryngoscopic Examination—December 9, 1865.—The mucous membrane of the fauces is much congested (presenting the appearance of raw beef), free border of the epiglottis and its posterior plane are of a rather subdued red color, the false vocal cords inflamed, and along the free border of the true vocal cords is a reddish-yellow margin resembling a seam.

The aryepiglottic folds as well as the mucous membrane covering the arytenoid cartilages have an unhealthy reddish-yellow appearance; phlegm is lodged on the walls of the upper part of the trachea. These appearances clearly account for the troublesome and persistent symptoms of pain, hoarseness and slight cough, and at times considerable expectoration.

Treatment.—The local application of nitrate of silver and the inhalation of atomized fluids brought about perfect recovery in two months. I have had several letters from the reverend gentleman since, in which he informs me that he has been able regularly to preach twice on the Sabbath, besides attending to his other duties, and when he made me a flying visit, last fall, his throat was in perfect condition. I ought to state, however, that I had demanded, as a sine quâ non, at the commencement of the treatment, his total relinquishment of public speaking whilst he was under my care.

CASE 2.—*Intense chronic catarrh of eight years' standing.*

Mr. R. H——, merchant, æt. about 30, well built and healthy, had been for years subject to a dry, hacking cough, which at times would constantly leave a dry feeling in the throat, accompanied by various degrees of hoarseness and considerable degree of expectoration. To these were added a sensation of fulness and dryness of the nose, and loss of sense of smell. Regularly every year, as winter approached, he grew worse, and had to seek refuge in the milder atmosphere of the South. Shortly before consulting me, in December 1865, he was advised by an eminent physician of this city to go again South, as nothing could benefit him save a change of climate.

Laryngoscopic Examination—Dec. 20, 1865.—The mucous membrane of the fauces is inflamed and irritable, covered with innumerable follicles over its entire extent, some quite large. The entire larynx, from the epiglottis down to the glottis, is of a dark-red color, resembling fresh meat. The epiglottis is thickened, inclined in position toward the arytenoid cartilages so as to cover fully one-half of the space between it and the former, and rendering it exceedingly difficult to get a good view of the vocal cords, which so far as visible appear inflamed. The mucous membrane covering the arytenoid cartilages is also highly inflamed and œdematous. The whole structure is irritable. Lungs in normal condition.

Treatment.—Inhalations of tar and sedatives to allay the irritation. Local applications of nitrate of silver, iodine and glycerine, daily for three months. One great difficulty experienced was, to bring the application to bear under the surface of the pendent epiglottis, a difficulty which, before the days of laryngoscopy, must have rendered many of the best directed efforts futile, in consequence of the remedy being applied to the outer anterior surface of the epiglottis, instead of to the inner posterior surface leading to the interior of the larynx. More often, the already much depressed epiglottis was rendered still more so, and sometimes no doubt to the great injury of the patient, by such well-intentioned but ill-directed applications.

The patient in this case recovered.

CASE 3.—*Acute catarrh of the epiglottis and true vocal cords.*

Miss V——, æt. 19, healthy, and an excellent vocalist, began to complain of hoarseness and dryness of the throat, in the fall of 1866, accompanied with a slight cough and pain in swallowing. This increased until she consulted me in October, and was compelled to discontinue her favorite pastime, viz., singing.

Laryngoscopic Examination.—External pressure upon the trachea and larynx produced no pain. The mucous membrane covering the epiglottis, and the contour of the vocal cords is of a deep-red color, strongly contrasting with the inner border of the vocal cords, which appear white, covered with some yellow substance. The same evidence of inflammation extends somewhat from the base of the epiglottis to both sides; for the rest, the larynx appears normal. Examination of the chest reveals traces of slight attack of bronchitis.

Treatment.—Absolute rest to the voice, local applications of astringents and inhalations. Recovery complete.

CASE 4.—*Intense catarrh of the larynx, with abscess near the tubercle of the epiglottis, opening toward the posterior or laryngeal surface.*

Mr. S——, æt. 35 to 40, stout and robust, of irregular habits, an inveterate smoker, presented himself May 1866, complaining of severe pain in his throat, aggravated by swallowing either fluids or solid food. Cough and complete loss of voice, which had continued for three weeks. Used gargles and medicines without effect.

Inspection of the throat.—The fauces are highly congested, the inflammation extending upward to the post-nasal region. The capillary vessels of the mucous membrane are very prominent and numerous, below which are seen innumerable yellow-grayish depressions resembling an ulcerated surface. The introduction of the laryngoscopic mirror barely tolerated. The epiglottis appeared much swollen, inclining toward its anterior or lingual surface. Below, near the tubercle of the epiglottis, is observed a considerable swelling, culminating into a distinct elevated yellow point, which we judged contained pus. The aryepiglottic ligaments or folds, the membrane covering the cartilages, false and true vocal cords, are inflamed. When the patient was ordered to pronounce the letter " æ," the vocal cords did not vibrate as usual—a fact which in the naturally large larynx of the patient could be perceived with great precision. The cough is troublesome but is confined to the bronchi.

Treatment.—Opened the epiglottic abscess with the requisite instrument, from which was discharged considerable pus, giving the patient almost instantaneous relief. Said the great pressure had gone. Ordered emollient gargles, poultices outside the throat, and an active cathartic. Next morning he felt much better, but the catarrahal inflammation, of course, continuing still severe, applied solution of nitrate of silver, gr. 40, to the ounce of water, to the fauces, as well as the larynx. In the morning and at night directed a sedative inhalation. Continued this treatment for nine days with the best results; the epiglottis was nearly normal, and the inflammation of the larynx much less severe; the red color of the vocal cords gradually disappeared. But on the tenth day of the treatment the patient still continued hoarse, the voice clear in the morning, but upon the least exertion hardly audible. Applied tinct. iodini comp. with glycerine, equal parts, for two weeks, with most satisfactory results. Voice was restored, and patient has had no relapse. An important part in the treatment of this case formed the administration of inhalations, which were invariably administered, when given warm, through Siegle-Bergson's apparatus.

2d. *The Sponge-Carrier*, as suggested by Prof. Richtor of Dresden and improved by Dr. Türck is designed for cauterizing the upper regions of the throat. In proportion as the piece of sponge fastened to the instrument is larger or smaller, the quantity of fluid introduced is regulated. The construction of the instrument is clearly discernible from the annexed cut and needs no further description.

Fig. 25 represents the Sponge-Carrier of Dr. Türck.

FIG. 25.

Türck's Sponge-Carrier.

3d. *The Syringe or Laryngeal Injector* has been adopted for the twofold purpose of introducing fluids into the larynx where it is impossible to do so in the usual manner, on account of great depression of the epiglottis, or some other cause, and secondly, for regulating the quantity of fluid to be injected by minims, such as the solutions of strychnine, when only one or a few minims are to be placed within the larynx by the aid of the laryngeal mirror.

Syringes are made of ivory, glass, silver and gold, graduated and non-graduated, according to the different purposes they are to serve. Tobold, Lewin, von Brüns, Gibb, Mackenzie (Rauchfuss' modification), and Türck, have each suggested valuable instruments. The last-mentioned ingenious gentleman has recommended a *sponge syringe*, particularly for children. The sponge syringe, of which the annexed cut is a representation, is made of india-rubber, with a curved, flexible silver tube, to the ring-shaped extremity of which

LARYNGOSCOPY AND RHINOSCOPY.

a piece of sponge is tightly fastened. Introduced into the larynx the fluid is injected.

FIG. 26.

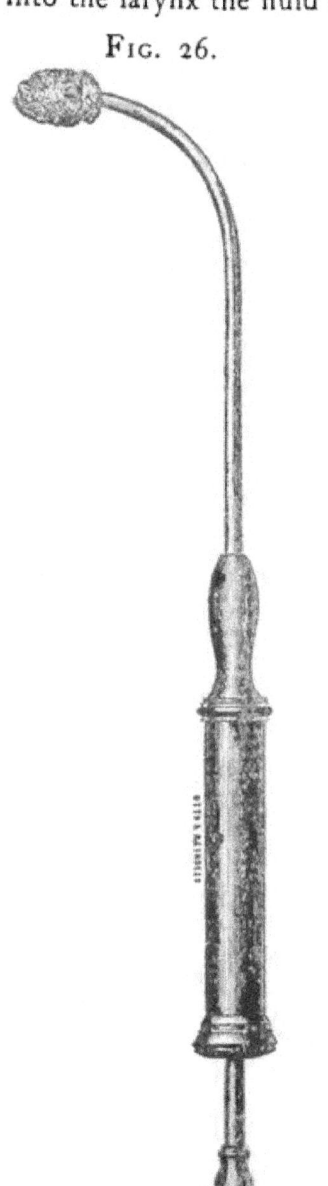

Türck's Sponge Syringe.

FIG. 27.

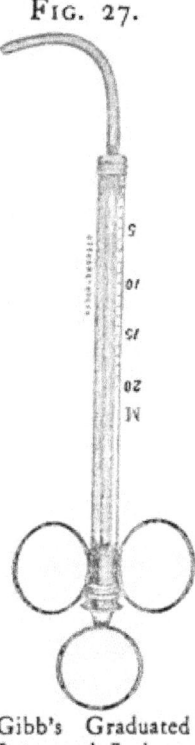

Gibb's Graduated Laryngeal Syringe.

Fig. 27 represents Gibb's Graduated Laryngeal Syringe.

Syringes are by far less often used than brushes, for the reason that the occasions already indicated not often happen to demand their employment, and as they have a tendency to cause more spasm than brushes.

4th. *Pulverisateurs,* or instruments for the application of liquids to the larynx, trachea, or bronchial tubes in the form of very fine spray, remain yet to be described.

The inhalation of various substances, such as the essential oils, sulphur, etc., by impregnating the vapor of hot water with them, is no new idea. Hippocrates, Galen, the Roman physicians, the Arabs, Piso in 1580, Bennet, a London physician during the middle of the 17th century and in our own time Trousseau, all have favored, more or less, inhalations in various forms of disease and had variable success.* Yet, as Dr. DaCosta justly remarks: "The exaggerated statements with reference to its action, the uncertain effects and the attempts to make inhalation serve the purpose of a panacea, produced again very naturally an utter want of confidence in them, which was only disturbed by the discovery in this century of iodine and chlorine."

Many eminent men, however, still in the memory of

---

* Our space does not permit us to enter at length into a discussion of the anatomization of liquids. We shall confine ourselves to what is essential for the student to know, referring those who desire to study this subject in extenso to the special works of Waldenburg, Lewin, Sales-Giron, Beigel, the reports of the French Academy of Medicine, Dr. DaCosta's compact and able monograph, as originally published in the *New York Medical Journal* for 1866; also Dr. Cohen's treatise on inhalation, 1867.

some of us, advocated the inhalation of the fumes of belladonna and stramonium, tar and turpentine, in bronchial affections. Still, great expectations of good to be derived from inhalations in consumption having been raised, and little realized in this very direction, the subject fell again into disrepute, and became the special monopoly of quacks, of which St. John Long was the prince;* representatives of whom infest communities to this very day ; for some of our fashionables, as well as more simple fellow-citizens delight to visit daily elegant establishments, where from flexible tubes extending in all directions, from apparatus most wonderfully constructed without regard to expense, they inhale the very breath of life and take a new lease.

As there were in times past, however, practitioners of good reputation who favored the practice of bringing solids and fluids in direct contact with the diseased membranes, so in our day, particularly since the development of the art of laryngoscopy, scientific men are not wanting who endeavor to place this important field of therapeutics upon a solid foundation.

The inhalation of atomized fluids in conjunction with a good knowledge of the use and application of the laryngoscope to the treatment of disease is destined, beyond a doubt, to accomplish much good ; but the habit or practice—now degenerating almost again into a *mania*—to manufacture inhaling apparatus for the million, to place one into every hand, for better or for worse, the only recommendation of which is its cheap-

---

* A Book about Doctors, by Jeaffreson, 1866, chapter St. John Long.

ness, and as a natural consequence its inferior workmanship, not fit to be used twice before being out of order, recommended by friends (and *physicians sometimes*) upon the fallacy that "it suited A's case, therefore it can do no harm to B. to try it," is, we fear, destined ere long to bring the whole subject again into disrepute among a large class of medical men and the public at large, and what is worse, to injure the well-directed efforts of those who make use of this therapeutic agent on purely scientific principles and after a carefully conducted diagnosis.

Anatomizers of fluids, or pulverisateurs, may be ranged under two classes:

1st. *Those where air acts as the forcing power;*
2d. *Those where steam is the motive power.*

The first, or the respiratory hydrostatic plan, as it has been called, was first advocated by Sales-Giron, by proposing a plan of breaking up fluid into very fine particles, and thus introducing these particles in the form of vapor into the larynx, trachea and bronchi. His first experiments were carried on at the thermal spring of Pierrefond, in France, where in a room set apart for the purpose, the mineral water was projected through a tube with great force by means of an air-pump placed near by. The fluid was forced from the tube through numerous capillary openings, and striking against the surface of a metallic disk, was broken up into vapor or mist, which the patients breathed together in the room.

The controversy which Sales-Giron's experiments first kindled, waxed warm in time, but the favorable report of the French Academy, opposition notwithstanding, and the fact that Giron's principle, viz., to drive a fluid by means of compressed air through a narrow opening to nebulize it, is adopted in all modifications of his first apparatus by others, establishes his claim to our acknowledgment of his genius. Auplan, at Lamotte les Baines, conducted similar experiments, but he caused the water directly to be dashed against the wall of the apartment, where it was pulverized much in the same way as a water-fall in striking against the rocks is broken into spray. Objectionable as the previous proceedings were, Sales-Giron soon successfully produced a portable, easily-managed apparatus for anatomizing fluids, which was exhibited at the Great Exhibition in London, 1862. It consisted of a glass vessel containing the liquid, to the neck of which an air-pump is attached. By pressing the piston the air in the interior of the vessel is compressed, and on turning the stop-cock it drives the fluid with such force against a metal plate contained in a barrel-shaped tube, that it is instantly converted into a fine mist, which the patient can easily inhale. The large tube, in the centre of the lower plane of which is placed a receiver, conveys away such portions as are at once condensed.

Various modifications of Sales-Giron's instruments have been proposed. Among others by Mathieu, Natanson, Bergson, Waldenburg, Weiss, all of which are good; but the best improvement is the apparatus of Lewin, which, from its simplicity of construction, and the facility with which it is kept clean, has had for years the preference in our practice, and is in daily use.

Fig. 28 represents Lewin's apparatus in operation.[*]

---

[*] Taken from Mackenzie's work on the Laryngoscope, p. 96.

Fig. 28.

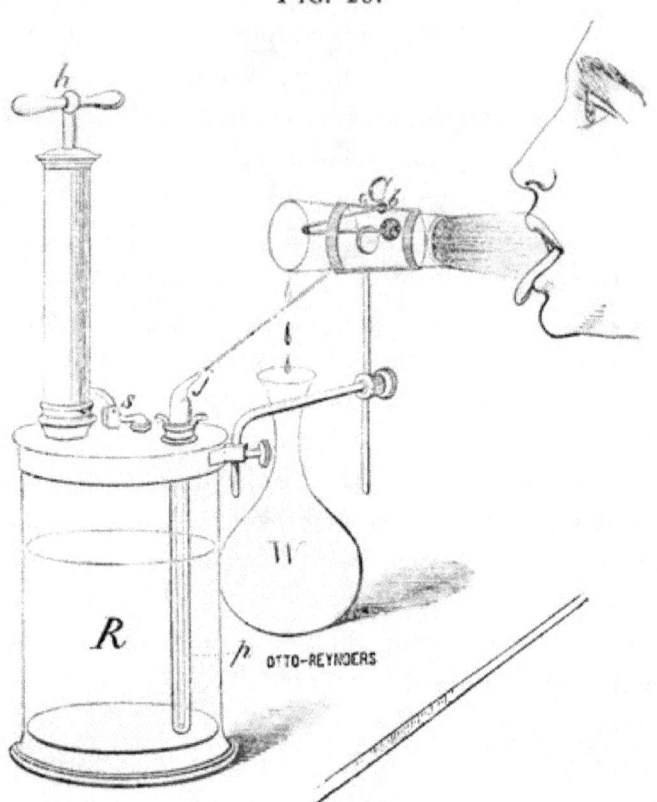

Lewin's Pulverisateur.

R. A glass receiver, into the metal top of which the air-pump is screwed. The inhaler is filled with the medicated solution by unscrewing the air-pump. Air is forced into the receiver by alternately depressing and raising the handle (*h*) with the right hand, whilst a finger of the left hand is kept on the extremity of the jet-thrower (*j*).

p. A fine glass pipe, which reaches almost to the bottom of the receiver, and after passing through the lid, is bent at an angle of about 130°. At its extremity is a fine opening.

j. The jet-thrower, through which a very fine stream passes to the metal button (*c*).

s. Safety-valve.

C. Glass cylinder for limiting the diffusion of the spray. It slants slightly, so that the further extremity is on rather a lower level than that near the mouth.

o. Opening in cylinder, through which the jet of liquid passes to

b. A metal button, on which the jet breaks into a fine spray. A portion of the liquid forms drops, which run into

W. The waste-bottle.

The patient's mouth should be placed close to the end of the cylinder, and the tongue protruded.

In the foregoing instrument the fluid, as has already been stated, is forced by air-pressure against some disk or other body, where it is broken into fine sprays and vapors, and then inhaled.

Mathieu, however, suggested an instrument which he called "Nephogène," afterward modified by Dr. Bergson, in which he uses the action of a current of air, compressed in a large ball, which, as it unites with the fluid, escapes with it from the capillary opening as a fine spray.

Dr. Bergson improved upon this process by placing two glass tubes, with capillary openings, at right angles to each other, in such a manner that the point of the vertical tube is close to, and about opposite the centre of the capillary opening of the horizontal tube. Whilst the vertical tube is dipped into the fluid to be pulverized, air is blown through the horizontal one. The air in the tube becoming rarefied, the liquid rises to the capillary opening and is there pulverized by the current of air from the vertical tube.

To avoid the labor and fatigue of keeping up a constant current of air by blowing into the tube, Dr. Andrew Clark substituted *the hand-ball anatomizer*, consisting of two balls, the lower of which is pressed by the hand, and the upper of which acts as an air-chamber.

The tubes of Bergson, made of best glass (though sometimes of metal), can be easily manufactured of all sizes, and with different curves, so as to pass up the nostril, or behind the velum of the palate, or into the larynx, and have only to be attached to Clark's rubber ball to make the apparatus complete. It is certainly a very simple, valuable and convenient instrument, and easily carried about.

Richardson's spray-producer for local anæsthesia, acts upon the same principle as Clark's and is the same apparatus slightly modified.

Dr. Gibb recommends an instrument which he calls *the laryngeal fluid pulverizer*, and which consists of a curved tube of silver, with an india-rubber receptacle at one end and a platinum capsule at the other, so finely perforated that the holes are invisible to the naked eye, yet permitting of the injection of a fine spray into the trachea throughout its entire length. For solutions too corrosive for silver, a similar gold instrument is used. Both instruments have rendered us good service several times.

Fig. 29 represents Gibb's laryngeal fluid pulveriser.

FIG. 29.

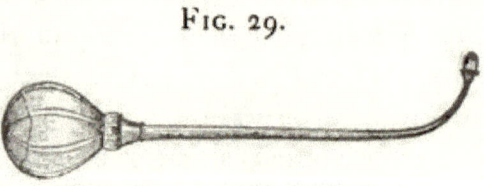

Gibb's Laryngeal Fluid Pulveriser.

Second.—The adaptation of steam as a motive power in the production of atomized fluids is due to Siegle of Strasbourg, France. He invented an ingenious apparatus, to which, adopting the arrangement of tubes of Bergson, he added a small boiler, in which steam is generated by means of a spirit-lamp or gas. The steam plays the part of compressed air, and as it escapes projects the liquid placed in the cup as a fine spray. A thermo-barometer was attached to the apparatus formerly, but is now considered unnecessary. The following figure (No. 30) represents Siegle's apparatus in its present most approved form, modified by Lewin, free from all unnecessary additions, and such as can be used in daily practice, without fear of its getting out of order, both by the physician and by the most uninitiated patient.

FIG. 30.

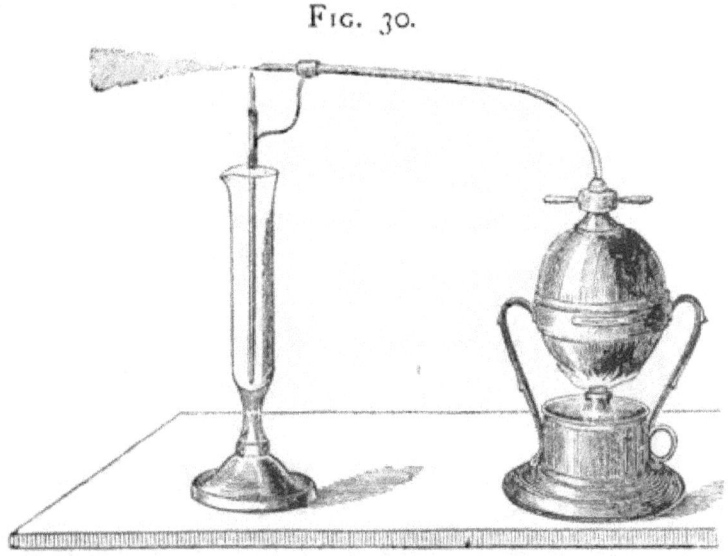

Siegle-Bergson—Lewin's Inhaling Apparatus.

The strongest testimony in favor of the usefulness and popularity of Siegle's instrument we have in the fact of its numerous modifications for sale in the market. But whatever may be said to the contrary, the apparatus, as represented in Fig. 30, has been but slightly if at all improved; in fact, most if not all modifications adopted are no improvements at all and in many cases are so many alterations for *the worse*. The *best* apparatus is the *cheapest* in the end; the *best* is *Siegle's*.

The Face Shield attached to some inhalers we consider superfluous, and may be dispensed with. It is cumbersome and annoying to some. It is only entailing another encumbrance upon the process of inhalation instead of rendering the operation as simple as possible. To avoid it, let the patient inhale through a glass funnel, of which the part introduced into the mouth is of convenient size, and the diameter of the opposite free part is from six to ten inches. The spray, as it escapes from the tube, passes through the funnel into the patient's throat, whilst the condensed portion drops into a glass standing before the person inhaling.

For the successful execution of the operation, a few observations may be useful. Whether the inhalation be taken warm or cold, it is important always to see that the tongue is not in the way of the current, so that the spray can readily reach the back of the throat.

Some of the patients press it readily against the floor of the mouth, some prefer to stretch it out during inhalation. Persons whose mucous membrane is sensitive, cough often during the act of inhalation. This may happen when the vapor is either warm or cold. It can be avoided by breathing very gently—not deep—for a few times, and if the inhalation is warm, to bring the funnel up as near as possible to the apparatus. Should this tendency to cough continue, the inhalation of simple steam for a few moments before the regular inhalation will overcome it.

The patient sits in a convenient position in front of the apparatus, so that the spray formed is on a level with the mouth, which is kept wide open and the head inclined backward. The distance to be occupied from the instrument varies according to circumstances. The proper distance to begin with is about six inches from the apparatus, which, as stated, may be increased or diminished as occasion demands.

Next to the breathing, its degree of intensity and velocity is to be regulated according as we desire to affect by the process either the fauces, or the larynx, or the bronchial tubes. The inhalation must never be continued so long as to be fatiguing, nor be taken on a full stomach. From ten to fifteen minutes is sufficient, and then the patient ought to rest, and not go out of doors immediately after, but be guided by the weather. It is impossible to lay down precise directions as to the frequency and time of inhalation. One inhalation per day is the minimum in a majority of cases; it is requisite to repeat the process two or three times. Instead of counting the number of inhalations, which is very fatiguing, the quantity of the medicinal fluid given to inhale is a reliable criterion, if properly regulated. As it is impossible for the patient to visit the doctor's office daily several times, it is best to provide him with an apparatus that is easily handled, readily kept clean, and reliable when set to work. And for these, no apparatus can better fulfil the above requirements than Siegle's, already described. For cold inhalations Lewin's pulverizer has, as already stated, our preference above all others. The patient must be properly instructed in the use of the instrument; he must see it taken apart, put together and set agoing. The medicinal agent must be carefully prepared and apportioned for him, lest he use too much or too little. The effects should be noted in a little memorandum-book kept for that purpose.

The above refers also to the solutions ordinarily used, of tar, tannin, sulphate of zinc and lead, alum, etc. When, however, the quantity of medicated fluid necessary is but small, as in some affections of the fauces and windpipe where solutions of caustic are required, it is recommended to use Clark's hand-ball atomizer, as better adapted to the purpose. Where corrosive solutions are to be introduced, Gibb's laryngeal fluid pulverizer will be found most useful.

The cases of affection of the throat and respiratory organs, in the treatment of which inhalations may be advantageously employed, are very numerous. To avoid tedious recital we have arranged for reference cases in tabular form as they have been treated by us, and their result so far as is known.

I.—Cases of General Congestion of the Mucous Membrane—Simple Catarrh.

No.	Name and Occupation.	Age.	Duration of Disease.	Length of Treatment.	The Agents Inhaled.	REMARKS.
1	Miss B. - - -	16	About 8 months.	Nov. 12th to Dec. 26th, 1865.	Aq. Piceæ; Tinct. Opii; Camph.; Zinci Sulph.; Ammon. Muriatis.	Health above the average; robust; tonsils much enlarged; chronic; removed by application of Vienna paste; congestion of the mucous membrane confined to fauces and upper part of larynx. Recovery complete.
2	Miss H. - - -	9	2 weeks, -	Nov. 17th to Nov. 24th, 1865.	Aq. Piceæ, Tinct. Opii.; Camph.; Alumen.	Subject to attacks of congestive sore-throat; anæmic and nervous; tonsils enlarged; reduced by Vienna paste. Recovered.
3	Master B. - -	11	9 weeks, -	Nov. 12th, 1865, to Feb. 7th, 1866.	Ammon. Muriat.; Alumen.	General relaxation of mucous membrane with congestion; scrofulous and weak. Recovered.
4	Mr. H., Merchant-	28	19 days, - -	Jan. 7th to Jan. 28th, 1866.	Ammon. Muriat.; Acidi. Tannici.	Had catarrh for several months, but only severe for the last fortnight; pain in the right side; cough at times quite severe; unable to attend to business; traces of bronchitis. Complete recovery. Had a relapse last winter, 1867. Recovered again.

I.—*Continued.*—CASES OF GENERAL CONGESTION, &c.—SIMPLE CATARRH.

No.	Name and Occupation.	Age.	Duration of Disease.	Length of Treatment.	The Agents Inhaled.	REMARKS.
5	Mrs. R., Seamstress	34	About 3 months,	April 19th to May 26th, 1866.	Aq. Piceæ, Extr. Hyoscyami.; Alumen.	Tuberculosis incipiens. Anæmic; both parents died of phthisis; had a severe cough which gradually diminished after the eighth day. Local application of tannin and glycerine to the throat. Recovered.
"6"	Mr. W., Clerk	43	7 months,	June 8th to Aug. 17th, 1866.	Ammon. Muriat.; Alumen.	Strong and good constitution, but careless; might have recovered sooner but for a second cold he took. Recovery complete.
7	Mr. S., Mechanic	41	16 months,	3½ months, Nov. 14th, 1866, to Mar. 2nd, 1867.	Aq. Piceæ.; Tinct. Opii.; Camph.; Ammon. Muriat.	Well built and good constitution. The cause of the disease is a neglected cold. Never took medicine, is averse to taking any. Is complaining of pain in the right side between the fourth and fifth rib when he gets fatigued. Recovered slowly, but has had no relapse.
8	Miss T.	8	18 months,	May 16th, 1867, to June 28th, 1867.	Aq. Piceæ.; Ammon. Muriat.	Congestion of mucous membrane with irritable cough. Recovered.
						N. B. In all these cases the larynx was examined as usual, and local application made to the mucous membrane with the brush.

II.—Chronic Catarrh, with and without Complications.

No.	Name and Occupation.	Age.	Duration of Disease.	Length of Treatment.	The Agents Inhaled.	REMARKS.
1	Mr. R. H., Merchant	28	7 years,	8 months.	Aq. Piceæ; Tinct. Opii.; Camph., Ammon. Muriat.; Zinci Sulph.;	Constitution moderately good; discharge from the nose and throat great; mucous membrane highly congested; follicles numerous; epiglottis depressed posteriorly; dry cough. Irritation and cough relieved by inhalations; congestion and follicles of mucous membrane by local applications of argent. nitr. Recovered.
2	Mrs. C.	26	3 years,	7 months.	Aq. Piceæ, Tinct. Hyoscyami; Alumen et Zinci Sulph.	Enlarged tonsils; removed by Vienna paste; incipient phthisis; constant discharge from the nose passing into the throat; larynx free from disease. Much improved.
3	Mr. K., Merchant	45	6 years,	11 weeks.	Alumen; Sodæ Muriat. et Tinct. Opii.	Elongated uvula—removed. Lungs weak and irritable; mucous membrane generally dry; cough dry and troublesome. Recovered.
4	Mr. B., Railroad Official,	32	Many years,	5 months.	Aq. Piceæ; Alumen et Zinci Sulph.; Ammon. Muriat.	Constitution good; mucous membrane of larynx highly congested and broken; no discharge of mucous; membrane dry and slimy as if painted; dry cough. Inhalations, applications of alumen et glycerine, etc., with brush. Recovery complete.

III.—CATARRH OF THE LARYNX AND BRONCHI.

No	Name and Occupation.	Age.	Duration of Disease.	Length of Treatment.	The Agents Inhaled.	REMARKS.
1	Mr. P., Clerk	29	Several years, but worse for two months.	July 18th, 1866, to Sept. 9th, 1866.	Soda Muriat. et Tinct. Opii.; Ammon. Muriat.; Alumen.	Periodically hoarse; sometimes entirely aphonic; mucous membrane of larynx congested; vocal cords relaxed and congested; symptoms of bronchial catarrh. Besides inhalations, local applications of alumen, tinct. iodini and glycerine. Improved.
2	Mr. R.	42	11 months.	August, 1866—Still under observation.	Ammon. Muriat.; Alumen.	Tickling and fullness of larynx; severe at times; very irritable; at times cough. Progressing satisfactorily.
3	Mrs. K.	29	2 years.	Nov. 19th, 1866, to Dec. 28th, 1866.	Alumen; Tannin.	Predisposed to phthisis; moderate inflammation of the laryngeal mucous membrane; nervous; hoarse and has pain when she wishes to speak. Local applications. Recovered.
4	Miss W.	18	4 months.	May 18th, 1867, to June 18th, 1867.	Alumen; Ammon. Muriat.	Dryness and pain in the throat; a little hoarse; considerable inflammation of laryngeal mucous membrane; also nasal catarrh. No cough; careless; gets readily cold. Left for Europe before cure completed, but much improved.
5	Miss M.	21	2 months.	June 11th, 1867, to July 28th, 1867.	Alumen et Zinci Sulph.; Ammon. Muriat.	Inflammation of mucous membrane of epiglottis and larynx as far down as true vocal cords; pain in talking; rough sensation; moderate expectoration. Much improved.

IV.—Syphilitic Ulcers of the Larynx and Pharynx.

No.	Name and Occupation.	Age.	Duration of Disease.	Length of Treatment.	The Agents Inhaled.	REMARKS.
1	Mr. R. - - - -	39	2 years, -	4 weeks. -	Tinct. Iodini comp.; Tinct. Opii.; Camph.; also internal administration of Hydrarg.; Bichlor. corros.; Gargles.	Syphilitic ulcer on the right wall of the pharynx; both tonsils much congested and enlarged; entire mucous membrane very red and covered with a thick, grayish-yellow secretion; vocal cords normal; ulcer touched with solid caustic. Recovered.
2	Mr. D. - - - -	43	9 weeks, -	2 months.—Still under treatment.	Tinct. Iodini. comp. Internal administration of Potass. Iodid. e. Syr. Sarsas. co. Gargles.	Mucous membrane much congested; tonsils swollen; exudation dirty gray; a large ulcer at the base of the uvula; pain in swallowing; condition much improved. Ulcer healed in 9 days.
3	Mr. T. - - - -	20	6-7 weeks, -	2 weeks. -	Hydrarg. Bichlor. corros. Internally, Potass. Iodid.; Gargles.	Pharynx and larynx much inflamed; hoarse, and pain in swallowing; plaques peculiar to the disease cover the pharynx; pain relieved after two inhalations; the patches disappeared gradually.
4	Mr. F. - - - -	49	4 months, -	11 days. -	Tinct. Iodini Comp.; also internal remedies as above.	Intense inflammation of the pharynx, larynx, both vocal cords; loss of voice and considerable cough; after the third inhalation the voice returned and pain ceased. Recovering.

THE PRINCIPLES AND PRACTICE OF

V.—ASTHMA.

No.	Name and Occupation.	Age.	Duration of Disease.	Length of Treatment.	The Agents Inhaled.	REMARKS.
1	Mr. M., Merchant	43	Many years.	From Dec. 1865, to April 2d, 1866.	Tinct. Lobelia; Tinct. Opii; Camph. cum Aq. Piceæ; Tinct. Hyoscyami.	Tendency to pulmonary disease; attack of asthma very severe at times, lasting for hours; inhalations would soon relieve the severity of the attack and check it. No cure, but great relief obtained at the outset of the attack.
2	Mr. R., Merchant	31	7 years	January, 1866, to May, 1866.	Aq. Piceæ; Tinct. Opii. Camph.; Tinct. Lobelia.	Asthma, dependent upon the condition of stomach chiefly; attacks are sudden and severe, but always relieved by the use of the inhalations.
3	Mr. N., Lawyer.	36	9 years	Only a week.	Tinct. Lobelia et Tinct. Opii. Camphor.	Confirmed asthmatic; tried the inhalation during two attacks with considerable relief. Lost sight of him.
4	Mrs. S.	33	11 years	Several weeks.	Tinct. Lobelia; Ammon. Muriat.	The asthma materially relieved during its continuance.
						N. B. Many more cases came under my observation. In all the patient was materially relieved of the suffering, although not permanently cured.

VI.—Tubercular Phthisis, Hæmoptysis.

No.	Name and Occupation.	Age.	Duration of Disease.	Length of Treatment.	The Agents Inhaled.	REMARKS.
1	Mr. K.	39	Unknown	3 months.	Aq. Piceæ; Ammon. Muriat.; Alumen.	Phthisis established; emaciated; considerable cough and expectoration; cough much relieved and expectoration diminished after using the inhalations for some weeks. Still employs the same.
2	Mr. L.	32	Unknown.	3 weeks.	Aq. Piceæ et Tinct. Opii. Camph.; Tinct. Ferri. Perchlorid.	Confirmed tuberculosis; large cavity in left lung; cough severe, but relieved by inhalation; repeated hæmoptysis stopped by inhalation of tinct. ferri-perchlorid.
3	Mrs. W.	39	Two years.	6 months.	Aq. Piceæ pura, et Ammon. Muriat.	Hereditary phthisis; dullness on percussion on both sides; depression under the left clavicle; severe cough; repeated slight hæmorrhages; distressed for breath day and night; inhalations gave complete temporary relief. Death from exhaustion.
4	Mrs. D.	23	7 months.	18 days.	Aq. Piceæ et Tinct. Opii. Camph.; Tinct. Ferri. Perchlorid.	Quick consumption; repeated hæmorrhages; cough obstinate; tinct. perchloride of iron checked it, as did also inhalations of salt water. Great relief at night from inhalation.
						N. B. Many other cases might be added.

Section II.

THE INTRODUCTION OF POWDERS INTO THE LARYNX.

Powdered substances, such as salts of mercury, zinc, copper, lead, alum, bismuth, or silver,* mixed in various proportions with very finely powdered sugar, may be introduced into the larynx by means of a brush, or by insufflation through a special instrument.

If the brush is used, which must be very fine, it is charged with the powder, as much as will attach to it, and then by the aid of the laryngeal mirror is introduced into the larynx.

The more reliable process is that of insufflation as already recommended by Trousseau and Belloc. Various instruments have been constructed for this purpose, such as a hard rubber tube, curved in front and provided with one or more holes. The powder is either introduced at the opposite end, or at a little opening in the tube, resembling a little window which can be firmly closed; or a portion of the tube may be unscrewed (Rauchfuss) for the purpose of introducing the powder into the straight piece of the instrument. The act of insufflation is accomplished by blowing either directly into the tube through a rubber

---

* The substances most generally used are nitrate of silver and tannin. It is well to begin with either in the proportion of 4 to 6 parts of sugar to one of the agents employed and gradually to ascend till the parts used are equal.

extension, the end of which is placed into the mouth (Störck), or through pressure upon a rubber ball, which is attached to the end of the hard rubber tube (Rauchfuss); or lastly, in order to prevent displacement of the tube during the expulsion of the powder, through pressure upon a rubber ball placed between the knees or under the feet, as advised by Czermak.

Section III.

THE APPLICATION OF SOLID CAUSTIC.

Caustics in solid form are preferable to the application of fluids or powders whenever it is desirable to cauterize powerfully a circumscribed point within the larynx or the surrounding parts, such as obstinate ulcerations at the root of the tongue, on the epiglottis, large granulations, excrescences and new-formations on either side of the free margin of the vocal cords, the base of growths after evulsion has been practiced. Preëminently above all stands the lapis causticus. Nevertheless, it is not without danger to introduce such an agent into the throat free, without protection, hence the numerous instruments devised to avoid the danger.

Dr. Störck of Vienna (who, when Laryngoscopy was yet in its infancy, in 1859, cauterized first with solids within the larynx), contrived a porte-caustique, in which the caustic remains concealed till brought to the part about to be touched, when it is made to protrude by pressure on a spring in the handle. Mackenzie has combined a guarded port-caustique with his admirable laryngeal lancet, by simply substituting a small piece of aluminum wire for the cutting-blade. The nitrate of silver is attached to the wire, which is bent at the same angle and of the same length above and below the angle as the laryngeal brush, roughened at its extremity and then dipped into some nitrate of silver fused over a spirit-lamp.* Tobold proposes a similar instrument, a bent silver wire, terminating into a solid flat point at the end, which is roughened with depressions so that the caustic will more readily adhere. Fauvel's porte-caustique has this peculiarity, that whilst the stick of nitrate of silver is safely inclosed, the point is always kept protruding by a spiral behind it. Rauchfuss, von Brüns, Tobold, Gibb, and particularly of late, Professor Türck, have proposed most useful porte-caustiques, each of which has its many good qualities to answer every purpose.

It is impossible sufficiently to impress upon the mind of beginners in this art the necessity of not undertaking the application of solid caustics to the larynx without having acquired the requisite dexterity of movement in the hand on the phantom.

After passing over the free border of the epiglottis it will be required to pass the instrument directly forward and down in the median line of it, provided we wish to cauterize at or near the anterior insertion of the vocal cords, and backward from the former point,

---

* Mackenzie, opus cit., p. 100.

if we desire to touch the posterior insertion of the vocal cords. If into the trachea, the movement must be swiftly performed during a deep inspiration. Such operations must be executed with the same ease as a catheter is introduced by a dexterous hand into the urethra.

The cough produced by these cauterizations is very severe generally—more so than after applying the brush; the pain produced is also sometimes very great and lasts for hours at times, which must not be overlooked but expected by the operator, who ought to forewarn his patient of the same.

Section IV.

ESCHAROTICS.

Escharotics are not often resorted to, except where the greater part of the mucous membrane of the larynx is covered with vegetations (which not frequently happens), and which it is useless to attempt to remove by the mouth, or as Ehrmann undertook to do by opening the larynx. The latter proceeding, if not foolish, at least promises but little success. According to Mackenzie the greatest benefit may result from the use of escharotics in such cases. Nitric and chromic acids, Vienna paste, and a mixture of caustic soda and lime are the agents, to the last mentioned of which Mackenzie gives special praise. The application of such remedies is however exceedingly dangerous and should only be undertaken by those who have had much experience in the introduction of instruments into the larynx.

To illustrate the satisfactory results of this application, the author quoted above relates a remarkable case where of five large, spongy excrescences in the larynx, viz.: one on the under surface of the epiglottis, another on the right ventricular band, a third on the left ventricular band, a fourth on the left vocal cord, and a fifth on the right vocal cord and the mucous membrane below the cord, the four first were removed with the forceps but the fifth on the right vocal cord and mucous membrane below, little effect being produced by the forceps upon this large growth, it was removed by the application of nitric and chromic acid in part, but the greatest benefit resulted from the employment of a mixture of caustic soda and lime. The growth was reduced to a quarter of its former size, and the patient recovered a loud and tolerably clear voice.

The following case happened in our own practice:

Mr. H——, æt. 26, of tolerable good health, consulted us in May 1866, about a choking sensation he felt at times in his throat, accompanied by partial hoarseness, which would last but a few minutes. This uncomfortable feeling would especially come on when driving his horses in the park, if he spoke suddenly loud to the animals, otherwise, he had no complaint. Upon laryngoscopic examination the mucous membrane of the fauces and larynx was singularly free from congestion. The epiglottis, which was large, was depressed so as to close up about one-third of the free contour of the larynx. On the left side of the free margin near the median line, was discovered a warty excrescence of the size of a small pea, resembling a cauliflower growth. With Prof. Türck's epiglottic pincette we raised the epiglottis, not without considerable difficulty, and applied chromic acid by means of the porte-caustique. We had the satisfaction of seeing the growth entirely disappear after four applications of the caustic. No return of the malady up to the present time.

CHAPTER III.

ON OPERATIVE PROCEEDINGS WITHIN THE LARYNX.

Section I.

Indications for operations: Scarifications and opening of abscesses.

ONE good result, remarks Dr. Johnson[*] in his admirable lectures, of the use of the laryngoscope will be, that henceforth fewer drugs will be consumed in cases of laryngeal disease, and those which are given will be administered with a more definite object and with a truer aim than heretofore. This in itself would be no slight gain, but the laryngoscope does more than this—it opens the way to methods of local treatment which

---

[*] Dr. Johnson. The Laryngoscope. 1864.

without its aid would have been impossible and inconceivable.

To this reference to operations within the larynx we may add, that as long, moreover, as we do not posess an anæsthetic which can be confined to the larynx and throat for a longer or shorter time, during which the operator remains within these organs with instruments, so long operations within the larynx must continue to be the most difficult upon the whole field of operative surgery. The delicacy and exquisite sensitiveness of the part we have to deal with, the circumscribed space within which our work is to be done, the danger which might result from some mishap, all these demand besides a sure, steady, light and practised hand of the operator, a union of favorable circumstances, which must be crowned by a most active coöperation and willingness on the part of the patient to do his or her share faithfully in this important work.

Hence a certain degree of education is necessary on the part of the patient, which is the work of days, sometimes of weeks. He must learn to open his mouth daily wider, to hold his head steady, and to bear not only the touch of the mirror in the fauces, but also the application of instruments within the larynx.

Operative interference is indicated :

1st. In cases of œdematous swelling of the mucous membrane, but particularly of the submucous cellular tissue within the interior of the larynx (œdema of the glottis), in consequence of or as a concomitant symptom of inflammation, suppuration or ulceration.

2nd. In circumscribed collections of pus, abscesses

in the submucous cellular tissue of the mucous membrane of the larynx, which in consequence of size and shape might encroach upon the free space of the larynx.

3d. In strictures within the larynx resulting from cicatrices of ulcerations, or in consequence of injury and loss of substance of the mucous membrane, from mechanical or chemical injuries, particularly in contraction of the glottis from scarifications.

4th. In the presence of morbid growths which may reach the size of a pea or a small cherry or more, and which Dr. von Brüns called cyst-polypi.

In such as the foregoing cases we resort, aided by the laryngoscope, to scarifications and incisions with remarkable success.

A great variety of instruments contrived and used by different operators—each claiming some advantage of construction over the other—has been constructed, all however having one necessary principle in common, namely, that of a small, double-edged knife or lancet which is contained in a hollow tube, suitably curved for introduction into the larynx. The point of the lancet is concealed in the duck-billed extremity of the tube, till forced out by pressure on a spring, either in the centre or at the extremity of the handle. On this principle is constructed the laryngeal lancet of Drs. M. Mackenzie, von Brüns, Winterich, and Tobold, with very slight unessential variations.

In Mackenzie's instrument the stock is provided with tubes bent at different angles, and below the angle is a joint by which to lengthen or shorten the tube. In the centre of the handle is a spring, which forces the lancet out when it is pressed down.

The advantage of this as well as all other instruments proposed by Mackenzie for operative purposes is this, that when the requisite dexterity in the use of one has once been acquired, all his other instruments, being constructed on the same principle as the aforementioned, can readily be applied to their respective purposes.

Von Brüns was the *first* who operated through the mouth with an open knife in the larynx and who extirpated thus a polypus. The stem to which the double-edged lancet was fastened was catheter shaped.

The tube of Tobold's lancet is slit open on each side. The knife is propelled by pressure upon a spring at the extremity of the handle. This instrument has many advantages, owing to the simplicity of its construction as well as to the ease with which it is handled. Reference to the same will be made in the following chapter.

Fig. 31, page 107, represents Prof. Türck's Laryngeal Lancet, with his latest improvements.

The extremity of the instrument is to be brought opposite the part which the operator wishes to lance or scarify, before he presses the finger on the spring. In œdema of the upper portion of the larynx as well as that of the glottis, the utmost care must be practiced not to make the incision too deep. It is safer to make repeated incisions at different intervals. The effect of the operation is in many cases almost instantaneous, and the benefit derived therefrom far exceeds the discomfort experienced during the same.

We shall give two interesting cases which came under observation lately.

LARYNGOSCOPY AND RHINOSCOPY. 107

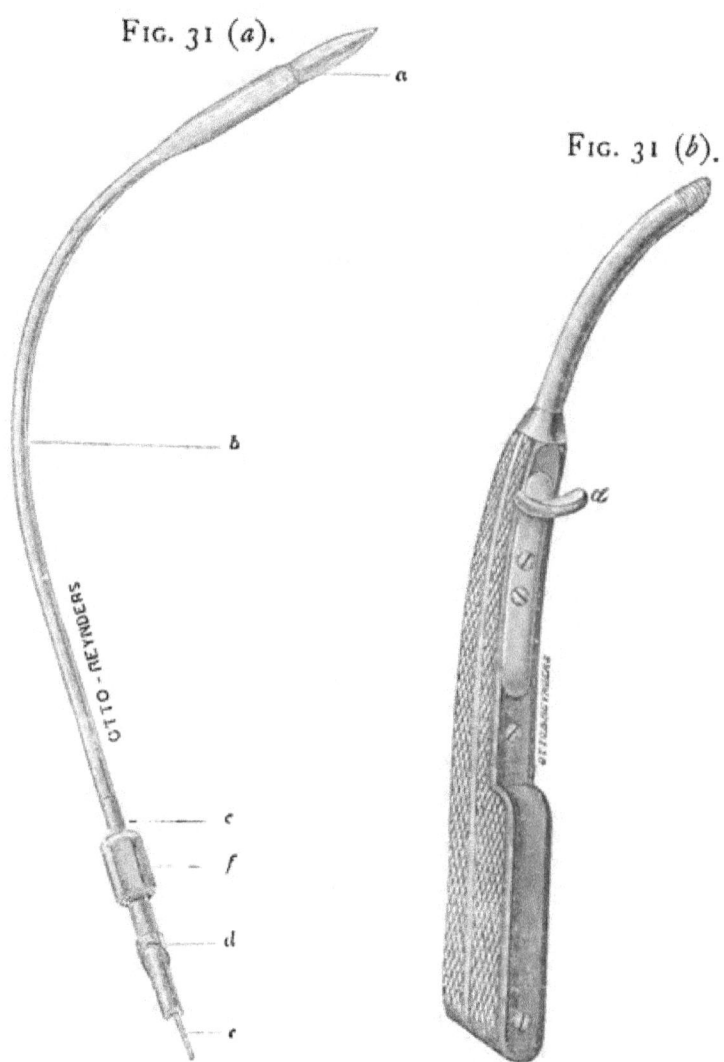

Fig. 31 (*a*). Large Laryngeal Lancet, natural size.

a, Exposed blade of the lancet; the posterior extremity broad, to prevent motion.

b. Part of the tube capable of being bent.

c. Part not to be bent.

d. Ring for uniting knife with handle.

e. Screw-like end of wire, which propels the lancet.

f. Screw, for the union of the handle with the other part.

Fig. 31 (*b*), represents the handle adapted to all of Türck's operative instruments yet to be described.

a. Shifter; within the tube of the handle is a second one which is moved by means of the shifter *a*, into which is screwed the end of the wire *e*. Fig. 31 (*a*). The union is the same as in catheters. In order to secure both more, screw *f*, Fig. 31 (*a*) is placed like a key over *e*. The tube and handle represent the segment of a circle, to facilitate the handling of the instrument.

CASE 1.—*Illustrations.*—*Œdema of the left ventricular band, causing great difficulty of breathing, hoarseness and pain. Cured by scarification.*

M. M., æt. 29., Irish servant at the Fifth Avenue Hotel, applied to me one morning in June 1866, complaining of great pain in the throat, hoarseness and difficulty of breathing. For five nights and days he had suffered frightfully, unable to lay down on account of the dyspnœa.

Laryngoscopic Examination.—The upper margin of the left ventricle and aryepiglottic fold formed a swelling which extended over one-half across the glottis. I could not see the left vocal cord. The color of the swelling was of a deep, rather dark red, but with a yellow spot in the centre. The mucous membrane over the neighboring structures was also inflamed and slightly swollen.

Treatment.—Lanced the part freely with Türck's instrument, which was followed by a quantity of blood. The next day, upon examination, the swelling had nearly entirely disappeared and the patient had slept well and had a good appetite. Made local applications to the part a few times with a solution of nitrate of silver. The cure was satisfactory. No return of the malady.

CASE 2.—*Abscess at the base of the epiglottis; great distress of breathing and pain; incision into the abscess; cure.*

Mr. A., æt. 32., an inveterate smoker and of irregular habits, consulted us in August 1866, on account, as he stated, of a sensation of choking in the throat. He could neither breathe nor swallow food without great pain and for two nights had no sleep, because he could not lay down, and had constantly to expectorate viscid matter. Used gargles and purgatives to no purpose.

The laryngoscope revealed the throat in a greatly congested state, the epiglottis swollen and much inflamed, whilst near the tubercle I could perceive a swelling of rather yellowish appearance, which corresponded to the point that the patient indicated from without with the fingers. At times the sensation felt was of a throbbing nature.

Active interference being decided upon, I introduced Mackenzie's laryngeal lancet by the aid of the laryngoscope, and opened by one incision the abscess, from which there was discharged about two teaspoonsful of matter and blood. Though suffering much, the patient felt much relieved at once, and in about half an hour after the operation was very comfortable. A few days sufficed for complete recovery.

Dr. Johnson also mentions one case in which puncturing the mucous membrane rapidly reduced an œdematous swelling over the arytenoid cartilages.

But one of the most splendid illustrations of the good which may result from scarification was the case of a boy with a cyst in the larnyx, which occurred in Guy's Hospital, London, in June, 1863, under the care of Dr. Wilks, operated upon and communicated by Mr. Durham to the Royal Medical and Chirurgical Society.[*]

---

[*] A description of the above case is, on account of its great value and interest, here appended, as given by Dr. Johnson, page 61, and reported in the Med. Times and Gazette of Nov. 21, 1863, and in the Lancet, vol. ii. p. 593, 1863.

"A boy, eleven years of age, was admitted into Guy's Hospital, under the care of Dr. Wilks, on June 10, 1863. He had for three years suffered from gradually increasing impairment of voice and difficulty of breathing and swallowing. On admission, all the symptoms were very severe. During the night of the 14th, he was seized, as he had previously been on several occasions, while asleep, with a very severe attack of dyspnœa. Tracheotomy was on the point of being performed, but was delayed by the desire of Dr. Wilks, and on the following morning Mr. Durham was requested to make a laryngoscopical examination. On doing so, the epiglottis could not be distinguished in its normal form, but instead there appeared a large, round, tense tumor, projecting backward and downward, and completely covering in and concealing the glottis; the tumor could be reached by the finger. Feeling certain that it contained fluid, Mr. Durham, with the concurrence of Dr. Wilks, incised it with a long, curved, sharp-pointed bistoury covered, except at its point, with sticking-plaster. The incision was followed by a sudden gush of thick, glairy mucous, mixed with a little pus and blood. All the patient's symptoms were at once relieved, and in the evening he was singing in bed. In the course of a few days he was perfectly well.

"When examined four months afterward, he was in every respect well. There was no appearance of the cyst, but the cicatrix of the incision was just visible on the lower part of the laryngeal aspect of the epiglottis."

Section II.

THE REMOVAL OF MORBID GROWTHS FROM THE LARYNX.

The greatest triumph in the treatment of disease by the aid of the laryngoscope has finally been achieved in the removal of tumors, polypi and warty growths from the interior of the larynx through the mouth.

Previous to the summer of 1861, when Prof. von Brüns of Tübingen, Germany, removed a polypus by decision, from the larynx of his own brother,* the extirpation of growths from the laryngeal cavity was of very rare occurence.†

---

* Die erste Ausrottung eines Polypen in dem Kehlkopf. Lihle, etc, by Doctor Victor v. Brüns, Tübingen, 1861.

† The following are the only well-authenticated cases of extirpation of growths from the larynx before the introduction of the laryngoscope, as collected by Dr. Mackenzie.

(a). Koderick, who operated successfully with a curved flexible tube, referred to by Semeleder, page 59.

(b). Pratt removed a tumor by subhyoid laryngotomy from the left half of the under surface of the epiglottis, which though it projected into the fauces, could not be got at from above. The tumor was firm, grayish-white, and fibrous. Semeleder, page 60.

(c). Sir Astley Cooper removed twice, with his finger, a large cancerous tumor about the size of a hen's egg, from the under surface of the epiglottis.

(d). Ehrmann removed a growth from the left vocal cord by first dividing the cricoid cartilage and several of the upper rings of the trachea; after 48 hours the larynx was divided in the median line up to the base of the hyoid bone. The parts being drawn apart the tumor was removed with the knife. Recovered in three weeks from the operation; the aphonia remained, but died of typhus fever five months latter. (Histoire. des Polypes du Larynx. Strasbourg. 1850).

(e). Dr. Horace Green, of New York, removed a pedunculated tumor (about the size of a cherry), which was (thought to be) attached to the left vocal cord. (Polypi of the Larynx, page 56. New York. 1852).

(f.) Professor Middeldropf, of Breslau, removed a tumor from the upper opening of the larynx by means of the galvano-caustic wire. Ruste, who saw the case six years

Since the introduction of the laryngoscope, however, these operations are by no means uncommon.

Professor Brüns, of Tübingen, who, as already stated, was the first to operate successfully on a polypus, aided by the light laryngoscopy had shed upon these classes of hitherto incurable diseases, removed 17 polypi or growths, alone, on as many different patients.[*] Lewin, Türck, Semeleder,[†] Tobold, Fauvel, on the continent, Dr. Walker of Peterborough, Drs. Gibb and Mackenzie, of London, Messrs. Bracey and Bolton of Birmingham, and undoubtedly others, have each, more or less often, extirpated growths from the larynx, with complete success. Dr. Elsberg,[‡] of New York, has also reported several cases. Three cases in our own practice we refrain, for the present, from reporting, as they are still under observation. They shall be made public at some future time.

---

after the operation, saw no symptom at that time of any return of the growth (Galvanocaustie W.).

If we except the first case, which is very vague, it appears that the growth could be seen in all cases which were operated upon with instruments, and in two instances (that of Sir Astley Cooper and Prof. Middeldropf) they could be felt with the finger. In the case of Ehrmann and Pratt the operations were indirect, and preceded by tracheotomy. "In Dr. Green's case," Dr. Mackenzie remarks, "the tumor could be seen, and though it was thought to be attached to the vocal cord, it more probably grew from the ventricular band or aryepiglottidean fold. If the polypus had been attached to the vocal cord, it could not have been seen projecting through the opening of the larynx, unless it had been unusually large, or its pedicle had been much longer than is usually the case."

[*] Dr. von Brüns, Laryngoscopische Chirurgie. Tübingen. 1865.
[†] Semeleder states the number of neoplasms removed since the introduction of the laryngoscope to be about 100.
[‡] Laryngoscopal Surgery, by Dr. Louis Elsberg. (Pamphlet. 1866.)

Dr. Sands, of New York, opened the larnyx for the removal of a tumor, in a case in St. Luke's Hospital. It projected from the left ventricle of the larnyx. Recovery complete, except that the voice remained a little rough. (Translator's note to Semeleder, p. 136.)

Dr. J. W. S. Gouley, of New York, has reported, very lately, in the September number of the New York Medical Journal, a case of polypus of the larnyx, and where Laryngo-tracheotomy was twice performed, the first time being followed by recurrence of the disease. (New York Medical Journal, for September, 1867, vol. v. No. vi.)

The special course to be adopted in removing a neoplasm from the larnyx is determined by its position, size and nature of insertion.

Sometimes one single particular operative proceeding suffices, oftener it requires two or more different methods.

A growth may then be removed

1*st*, *By the operation of Cutting : Decision, Excision and Puncture.*

2*nd*, *By Crushing.*

3*rd*, *By Cauterization.*

4*th*, *By Galvano-caustic.*

1*st. Decision, Excision and Puncture.*—These operative proceedings deserve first consideration according to Tobold,* wherever there does not exist, from some cause or other, an absolute impossibility to enter the cavity of the larynx with sharp instruments. Besides the character of the growth itself, the size of the pharynx and larynx are important factors entering into consideration.

Polypi, attached either by a thin or strong stem to the lateral walls of the larynx above the vocal cords, in a horizontal position, as well as those inserted into either free border of the glottis and hanging into the cavity of the larynx are to be operated upon either by *decision*,† that is by making repeated incisions into the growth, after which, in the course of two to four days generally, it is thrown off as a disintegral mass, leaving a little suppurating stump which is usually cicatrized in a few days more, or by *excision*, which implies the total removal of the polypus at its base, the complete-

---

* Die Chronischen Kehlkopfskrankheiten. Dr. A. Tobold. Berlin. 1866.
† First practiced by von Brüns with the knife. 1861.

ness of which operation will depend upon its attachment, and lastly, by *puncture*, which modus operandi is employed for the destruction of *cyst polypi*, when after puncture the colloid mass or fluid contained within the cyst oozes out and the growth disappears.

Numerous, almost too numerous to be mentioned, are the instruments employed for the above purposes. Those of Dr. Tobold claim here a prominent place on account of their usefulness.

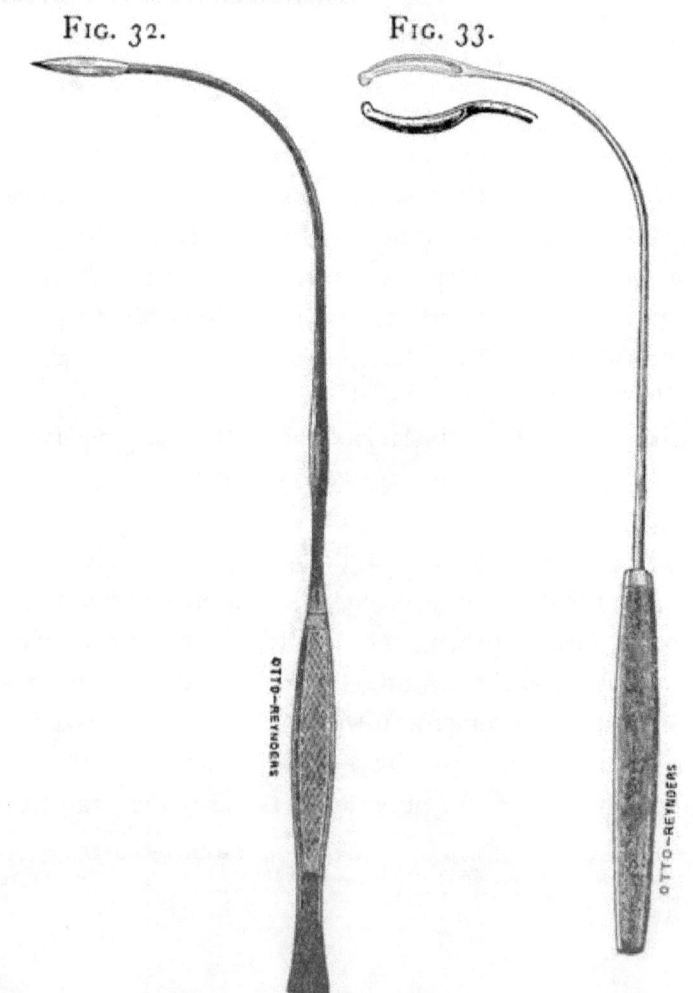

Fig. 32.   Fig. 33.

LARYNGOSCOPY AND RHINOSCOPY. 115

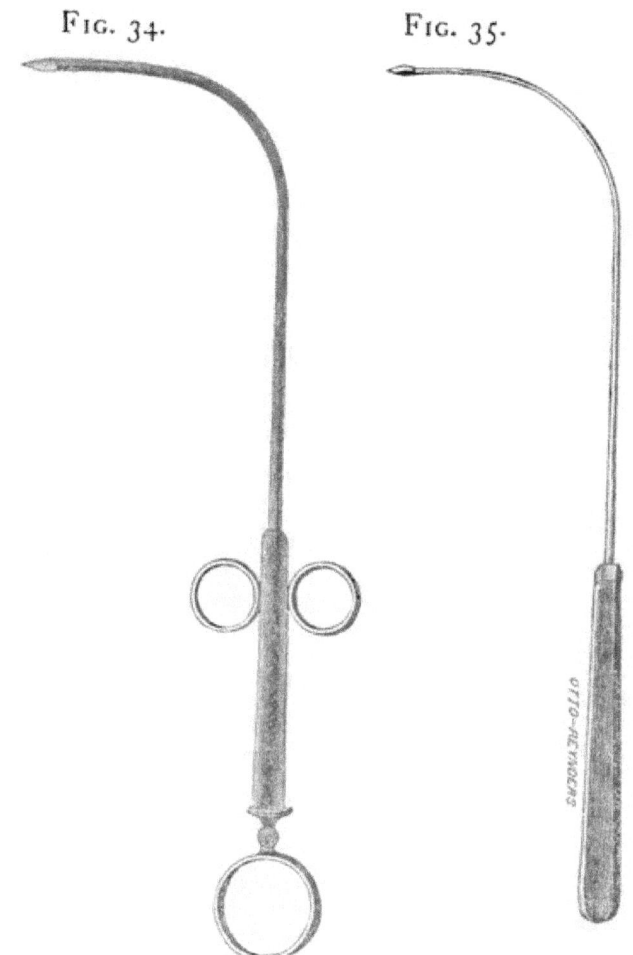

Fig. 34. Fig. 35.

Figures 32, 33, 34 and 35 represent the instruments used by Dr. Tobold for decision, excision, and puncture.

Voltolini has suggested an instrument similiar to the tonsil-guillotine of Fahnenstock.

Prof. Türck* has of late introduced several ingenious and useful instruments, which are here for the first time brought to the notice of the profession, except in the original edition of his superb work lately published.

---

* Opus cit., p. 573.

116   THE PRINCIPLES AND PRACTICE OF

Fig. 36.   Fig. 37.

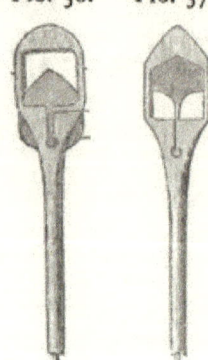

Fig. 36, and Fig. 37, represent what Türck calls his large fenestrated knives, similar to a tonsil-instrument, handled in like manner.

Fig. 36 represents the instrument open before the operation, Fig. 37, the same closed after excision. It is very flat, both sheath and knife very thin, must be held firmly in its position during the operation and can be used to extirpate hard or soft substances with either broad or narrow attachments.

Fig. 40.

Fig. 38.   Fig. 39.

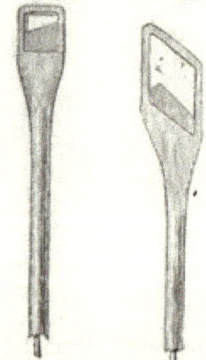

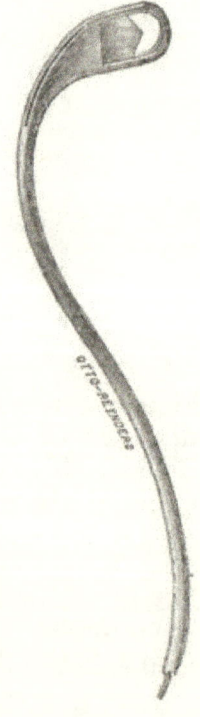

The operation can only be successful, if these instruments are used, when we succeed to secure the growth into the same just as a lasso is thrown around an object.

Figures 38, 39 and 40 represent smaller fenestrated knives, different in construction, of which Fig. 38 is specially applicable to the removal of growths from the inner borders and under surface of the true vocal cords, Fig. 39, for bodies near the anterior angle of the glottis, and if the stem of the instrument is modified by being properly bent, the same knives can also be used for other operations in the throat above the larynx.

Fig. 40 is applicable for the removal of neoplasms from the inner plane of the aryepiglottic folds, on the upper surface of the false and possibly also true vocal cords.

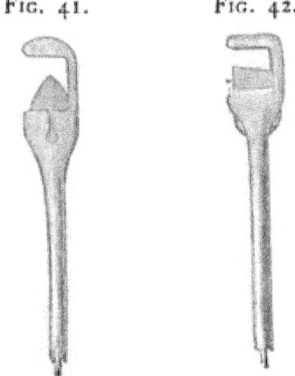

Fig. 41. Fig. 42.

Figs. 41 and 42 represent knives without frame on one side, applicable for new formations too large to be inclosed within the frame. The knife must be adapted to each case.

Figs. 43 and 44 are what Türck calls his sheath-knives, made like his instrument for crushing polypi, with the exception that the lower arm or blade is provided with a sheath for the reception of the upper sharp cutting one.

Fig. 43.   Fig. 44.

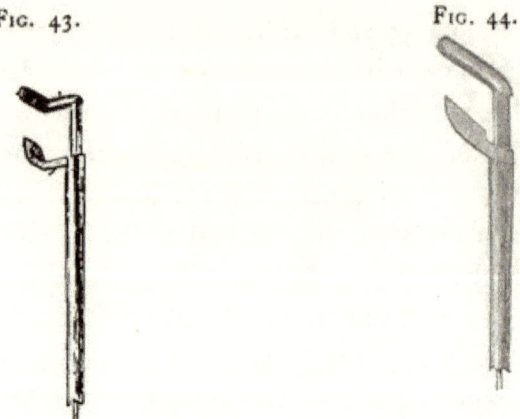

In Fig. 43 the blades extend obliquely from a second short arm, capable of being either placed toward the handle of the instrument, or vice versâ. If directed toward the handle, it is adapted for formations arising from the posterior surface of the epiglottis, if forward, for the anterior surface of the same or for the anterior plane of the posterior wall of the larynx.

It remains yet here to mention the scissors-shaped instruments for the above-mentioned operations, though perhaps not so frequently used as formerly.

When the larynx is large and all other conditions favorable, these complicated instruments are to be recommended and may be used to advantage; but these desirable features being wanting and the throat and larynx only of ordinary size, they are objectionable on account of the space they occupy within the larynx, interfering considerably with the complete view of the parts and place of operation, whence it it is not unfrequently impossible to see in the mirror the part of the instrument which actually accomplishes the process of cutting or to keep the same in view during the progress of the operation.

2nd. *Crushing* is another mode of destroying intralaryngeal morbid growths, whereby their nutrition is impaired and mortification is produced.

This process of procedure can only be applied to such neoplasms as are free, that is, unattached except by a thin feeble stem and which, either singly, or in a mass, are suspended within the larynx.

When, on the contrary, the mass to be extirpated is extensive and of soft texture, conditions which are not unfrequently met with, or cannot be reached by any other means, a modification of the process of crushing, namely that of *evulsion*, has been proposed and in some cases successfully executed, though deprecated by Tobold and other observers.

The operation is executed with pincers and forceps (polypus-crushers), of which those proposed by von Brüns, Tobold, Moura-Bourouillion, Semeleder, Leiter and Winterich are excellent. *Pincers* of various construction have been used for the operation, yet they are inferior to other instruments in regard to strength and action. They are useful for extirpating soft neoplasms when in a horizontal position, as compared with their point of attachment.

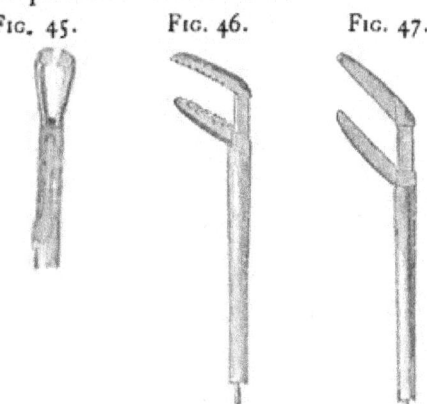

FIG. 45.   FIG. 46.   FIG. 47.

FIG. 48.   FIG. 49.   FIG. 50.   FIG. 51.

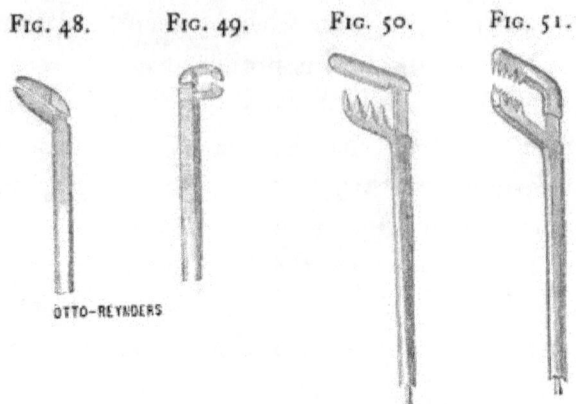

OTTO-REYNDERS

Fig. 45, gives Türck's pincette, adapted to soft growths.

Figs. 46, 47, 48, 49, 50, 51, represent the various polypus-crushers or polypus-forceps of Professor Türck.

Fig. 46, Large doubly indentated polypus-crusher.
" 47, Large sharp polypus-crusher.
" 48, Small sharp polypus-crusher.
" 49, Sharp polypus-crusher, with blades placed square.
" 50, Single indentated polypus-crusher.
" 51, Large double indentated polypus-crusher.

Each of these instruments is to be chosen with reference to the peculiarities of the case. As a general rule the indentated forceps or crushers are well adapted for soft masses, and the sharp double-bladed instruments for hard, resisting growths.

An attentive examination of the foregoing cuts will readily show wherein the difference and advantage of each respective instrument exists.

For a more detailed description, the reader is referred to Türck's Clinic on the Diseases of the Larynx,

etc., where each instrument is separately described.*
Suffice it to say, that all of the many difficult operations within the larynx performed by this most ingenious and successful laryngoscopist are accomplished by the aid of these instruments.

Ligation, which comprises the removal of laryngeal growths by the *galvano-caustic ligature* and by the *ecraseur*, is simply a modification of the operation of *crushing*, and is properly considered here.

Galvano-cautery forms the subject of a subsequent chapter; the *ecraseur* demands now our attention. The object of the ecraseur is, to catch with a loop of fine wire the laryngeal tumour like a noose, to draw it home (tight), when the pedicle is cut across and the tumour detached. Dr. Gibb, of London, first suggested the use of the ecraseur and his first instrument was constructed after Wilde's snare, used for the extirpation of aural tumours. The wire loop projects from the extremity of a curved tube, which is divided into halves, by a partition running through it. The free ends of the wire project from the openings of the tube behind, and are wound around a cross-beam which slides on the bar or tube, against which the fore and middle fingers rest.

FIG. 52.

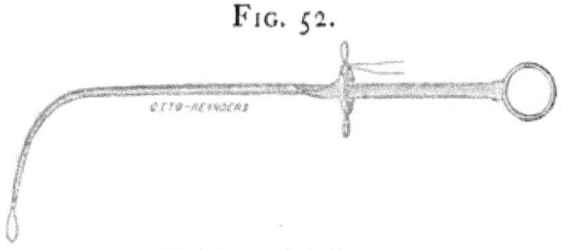

Gibb's Laryngeal Ecraseur.

* Opus cit., p. 570, et seq.

Fig. 52 represents Gibb's improved laryngeal ecraseur. Windler, of Berlin, has also constructed a most admirable instrument, similar to the above.

This operation is preferable for the removal of soft tumours, with a thin, stem-like attachment; it can, if desired, be replaced by the use of the above-mentioned instruments.

3d. *Cauterization* having already been considered (Part ii., Chap. 2, page 67, under Section i., Solutions, Section ii., Powders, Section iii., Solid Caustic, Section iv., Escharotics), the student is referred to these respective sections for information on this subject.

We may add here, that cauterization, especially with solids, is serviceable in removing small neoplasms. After operative extraction of growths, especially those of the character of papillomæ, the extent of the attachment at the base may be cauterized, in order to prevent a re-formation of the same abnormal excrescenses.

4th. *Galvano* or *Electric-Cautery*. Correct as the reasoning is, that *galvano-cautery* or *galvanic heat*, as it is often called, can be applied to the destruction of growths within the larynx, in the same manner as has been done for a long time, with advantage, in cavities of the body accessible from without, yet no such successful attempt had been made before Dr. von Brüns applied this agent in two cases of polypi of the larynx in 1864;* although recommended by Czermak, and later practised by Voltolini.

---

* The case reported by Middeldorpf (Galvano-Caustic, contribution to Operative Medicine, Breslau, page 212, et seq.), cannot be considered an operation of this kind

According to Dr. von Brüns* the difficulty which we encounter in the application of the galvano-cautery in practical laryngoscopy arises first of all from the want of proper instruments adapted to the operation.

Two objects have to be attained: first, the instrument must be so constructed as to occupy as little space as possible, and secondly, the wires or conductors must not only be of a certain definite size up to the platinum-chain, but these must remain perfectly *isolated* from one another, else upon the chain being closed, instead of the platinum-wire alone being heated to glowing, the entire chain becomes simultaneously heated. A further drawback will be found in the management of these instruments, arising from the connection of the extremity of the conducting wires with the battery. These wires, in order to be good conductors, must be of considerable size and solidity, which interferes very much with the necessary (sometimes almost imperceptible) movements of the free extremity of the galvano-cautery. Hence this resistance of the wires must be overcome by greater exertion of force, which interferes considerably with the motions of the sensitive fingers through which that very force has to be exerted. No operation of this nature ought, therefore,

---

within the cavity of the larynx. He removed a tumour that could be seen and felt in the pharynx, which arose above the superior thyro-arytenoid ligament, from which it grew upward into the upper part of the laryngeal cavity. The galvano-caustic chain was applied with the finger, and the part above, as deep as the aryepiglottic ligament, was entirely removed, so that fifteen minutes after the operation the stem of the tumour, of the size of the little finger, could be felt on the upper aperture of the larynx. Hence an operation not within the cavity but directly above.

* Opus. cit., 244, et seq.

to be undertaken without an assistant, who holds the two wires and endeavors to guide them according to the motions and requirements of the operator.

A final and not unimportant consideration is, that the galvano-cautery, in order to accomplish its purpose, has to remain longer within the larynx, in contact with the object to be destroyed, than is the case when knives or ecraseurs are used. The chain must remain applied at least three, four, or five seconds, until the purpose is accomplished. When, however, the part with which the chain comes in contact is only in the least sensitive, an inclination to cough, often severe cough, follows immediately the application, or at least after two or three seconds, when we are obliged instantly to discontinue the application and remove the instruments.

If, on the contrary, when the cautery is well borne and in full force, where it has been applied, contraction of the walls of the larynx takes place, it happens not unfrequently that the opposite wall is touched by the cautery, in the same manner as happens during the application of caustic solutions. If proper care is exercised no bad effects can follow from this mischief in the end.

Owing to the severity and rapidity of its action, galvano-cautery is with difficulty applied to a small circumscribed superficial spot, as is readily done with solid caustic. Suppuration, following the application, is also more extensive in the former than the latter application.

It follows then, that as advantageous an agent as galvano-cautery is for the entire destruction of morbid growths of considerable size within the larynx, the removal of which on account of form, nature, etc., is rendered impossible in any other way, it is not applicable to the extirpation of small neoplasms situated on the vocal cords.

No known cautery is so powerful, yet its work is done without blood, particularly if the platinum-wire is but gently heated, while the pain caused by the operation is hardly felt.

The last and most indispensable condition for the success of the operation requires, that the application of the platinum-wire or chain to the part to be destroyed be distinctly seen and watched by the operator.

Of the instruments for galvano-cautery, among the few that have been constructed, those of Brüns are the most serviceable and simple. They ought not to be used for the purposes above indicated, except in the hands of a ready and practised laryngoscopist. In such hands, from the very fact of the exceptional cases in which it is employed and the difficulty of its application, laryngoscopal surgery will achieve its greatest triumph. (von Brüns.)

Very few cases have thus far been reported whereon Galvano-Cautery has been successfully applied. The difficulty of the subject has undoubtedly deterred many from using it, besides the rare opportunity which is but granted to few in practise, to have under treatment such cases as require the application of this agent. When more numerous data shall be at hand from different observers, it will then be possible to compare the result of this operative procedure with others employed under similar circumstances.

CHAPTER IV.

THE APPLICATION OF GALVANISM TO THE LARYNX IN TEMPORARY OR PERMANENT APHONIA.

The deviations of the human voice from its normal condition are so numerous and subject to so many causes, that the study of the rational treatment of these has received a new impulse with the era of the laryngoscope.

In the language of Dr. Stokes:[*] "The field is open and promises a rich harvest." The field has been entered by many observers. That wonderful and beautiful mechanism which gives rise to phonation is now capable of being studied in its healthy as well as in its diseased conditions. The various complaints classed under the name of "*Aphonia*," have to yield to the well directed efforts of the hand guided by sight.

In discussing the application of galvanism to the treatment of aphonia, exception is made to *organic* aphonia, arising from inflammation, whether acute or chronic, induration or thickening, œdema above or below the glottis, ulceration, growths and tumours, or disease of the brain.

On the other hand galvanism deals with functional aphonia, which is manifested by certain modifications of the voice, that may vary from inconsiderable hoarseness, (partial aphonia), to that of a total loss of voice, (total aphonia).

---

[*] Diseases of the Chest.

To illustrate this more clearly, the subject, instead of having the power to give utterance to sounds at the same pitch, high or low or both, as before, can utter or sing only a few notes of the register or scale, peculiarly circumscribed (oligotonic), or he may be able to sound a single note (monotonic), or the register of his voice may be higher or lower than before, in the latter case often acquiring a peculiar *timbre*, as for example the *falzetto voice* of some artists; or the voice, in addition to its register becoming circumscribed, sounds harsh, hoarse and fatiguing, its purity and strength gradually diminishes, till in the place of the former sonorous tones we only hear a gentle feeble effort to speak.

This is evidence manifested to us through the organ of hearing of the presence of aphonia or phonetic paralysis, to which we must now add the evidence afforded us by the laryngoscope. Through the latter we learn, (having excepted all cases of organic aphonia as above), that the causes of the various stages of aphonia, both temporary or permanent, are owing to either a partial or complete paralytic condition of the muscles of tension and motion of the vocal cords and arytenoid cartilages, more directly shown in the greater or less degree of perfect approximation of the vocal cords.

The cause of this abnormal condition must, therefore, be situated either in the muscles which immediately govern the degree of tension and motion of the vocal cords, inasmuch as these muscles have lost the power of receiving the nerve-action necessary to produce contraction, or the power to react promptly to the nerve-influence; or the cause may be inherent in the nerves supplying these muscles, particularly the inferior and superior laryngeal nerves, the power of which, to conduct the cerebral impulse necessary to speech to the respective muscles, is either diminished or weakened, or entirely suspended. To one or both of these abnormal conditions we may trace back all those cases of functional aphonia as may arise from the emotions, such as joy, anger or fright, from hysteria, impaired innervation, certain local influences, congestions and strains, exhausting diseases, pressure on nerve-trunks, poisons, as the narcotics lead, antimony, and arsenic.

To remedy such conditions as enumerated above, the application of the galvanic current is rationally indicated as one of the most powerful means to stimulate and restore the action of nerves and muscles.

For this purpose the *inducted* current (more rarely the constant current) is used. The extremity of the poles attached by the wires to the proper apparatus is olive-shaped, as advised by Duchenne. Dr. Mackenzie has substituted an improved laryngeal galvanizer, in which the current does not pass beyond the handle, till the sponge is in contact with the vocal cords, the handle is pressed down at the proper moment by the index finger.

This admirable instrument has supplanted nearly all others. Being constructed on the same principle as all of Mackenzie's instruments, its use will be readily acquired and its advantages appreciated. We use this instrument with one of Stöhrer's (of Dresden) coal and zinc batteries.

Fig. 53 represents Mackenzie's Laryngeal Galvanizer.

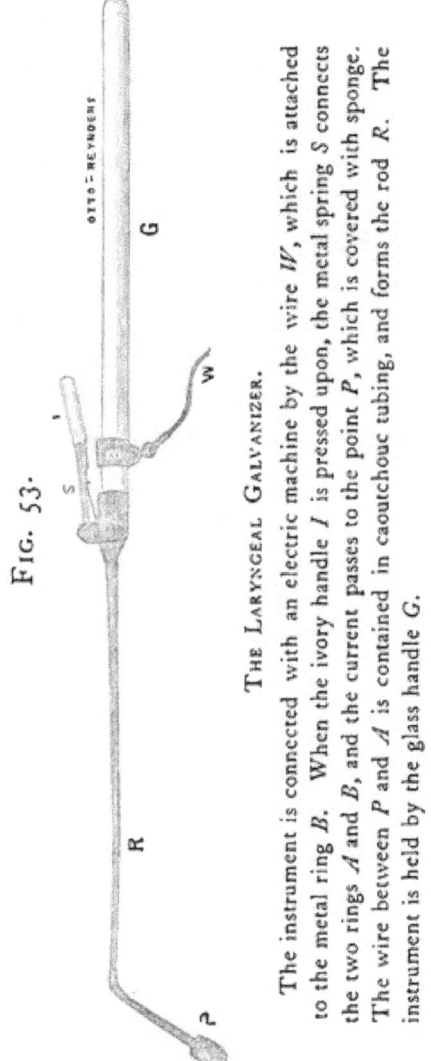

Fig. 53.

THE LARYNGEAL GALVANIZER.

The instrument is connected with an electric machine by the wire W, which is attached to the metal ring B. When the ivory handle I is pressed upon, the metal spring S connects the two rings A and B, and the current passes to the point P, which is covered with sponge. The wire between P and A is contained in caoutchouc tubing, and forms the rod R. The instrument is held by the glass handle G.

It is held in the hand between the thumb and second finger; when the sponge is in contact with the vocal cords, the operator presses with his index-finger on the spring in the handle and the electric current passes through the larynx to the skin externally.* To facilitate the application to the vocal cords, Mackenzie suggests to the patient the wearing of a kind of elastic necklet, in the centre of which is a piece of metal covered with sponge. This plate of metal, which is inclosed in cotton, is about three inches long and one and a half broad, and is bent back in the centre, so that, when applied, it corresponds to the thyroid cartilage. Projecting forward from the centre of this thyroid pad is a metal eye, by which it may be connected with the electric machine. The pad should be wetted before it is put on the patient's neck. The employment of this necklet enables the operator to dispense with assistants. When the point of the galvanizer is placed on the vocal cords, the electric current passes right through them in all directions, to reach the pole over the thyroid cartilage.

Drs. Smyley, Johnson, Tobold, Fauvel and many others have borne testimony to the value of this instrument and we cannot but join gladly in their commendations. We use it in preference to any other of its kind, with and without the necklet.

It remains now to be stated where the current is to be applied.

1st. Both poles of the battery may be directly applied to the mucous membrane of the larynx, being placed in close proximity to one another in the space selected for the purpose.

---

* Mackenzie, page 101, c. 103.

2d. One pole only may be placed upon the mucous membrane of the larynx, whilst the other is applied to the skin covering the larynx from without; thus, by placing the sponge of the galvanizer on the arytenoid cartilage, both branches of the pneumogastric nerve are stimulated.

3d. Both poles can be brought in contact with the throat externally, as for example in paralysis of both vocal cords and appendages, where one pole is to be applied on each side of the larynx, in order that the current may pass obliquely through the larynx and neck, or both poles may be applied on the same side of the throat, one near the junction of the clavicle with the sternum, the other above, on the side of the larynx, in order that the current may progress from below upward, to answer the course of the recurrent branch of the pneumogastric nerve.

Whether the preference should be given to percutaneous application of galvanism over the direct action of the current upon the mucous membrane, or vice versâ, is a point which only carefully collected and digested data of cases can decide, although certain it is, that the direct application of the current to the mucous membrane has in some cases produced results not attained by the opposite method.

Could we exactly know, in every case of nervous aphonia, for example, which muscles refuse to act, and what branches of the inferior or superior laryngeal nerve are implicated, it would then be easy, by the aid of the laryngoscope, to act directly and almost exclusively upon these muscles and nerves. Direct faradization of each separate muscle, or part of one, and each distinct nerve-ramification, remains therefore as yet a problem to be solved.

The direct operation is simple. If it is intended to let one pole act upon the mucous membrane, the patient is directed to take the same position which he assumed for a laryngoscopic examination, whilst he at the same time presses with his right hand the other pole externally against the spot indicated by the operator. The physician himself introduces with his left hand the mirror, with the right hand the instrument described before, held like the brush or laryngeal lancet when introduced into the larynx. As soon as the pole reaches the spot selected for the application, the little ivory handle in the centre is pressed down, when immediately the action of the galvanic current is manifested, according to its degree of strength, by sensations more or less acute, prickly, sharp or burning and contractions of the muscles brought under its influence. Should the sensations and contractions be too severe, then the current is to be interrupted for a few moments.

When both poles of the battery are to be applied to the laryngeal mucous membrane, the greatest care must be exercised in their position upon the respective muscles. Preliminary attempts ought first to be made without the poles being attached to the battery.

The duration for each application varies from one-quarter to one-half to three-quarters of an hour, generally once a day, except in such cases where its effect again rapidly passes off, when the operation may be repeated two or three times daily, but then only when the current is applied externally. The time required for a perfect cure varies from two to four to six weeks, sometimes longer. Few cases are reported where decided amelioration has taken place after the first application and the improvement remained permanent.

In nineteen cases of functional aphonia which have come under our observation, eleven of which were over one year's standing, and the balance below that time, we employed Mackenzie's galvanizer; twelve cases were entirely cured, and seven materially improved. The average time of each sitting was twenty-five minutes, repeated daily, and the average time required for treatment about twenty days.*

CHAPTER V.

GYMNASTIC OF THE LARYNX IN APHONIA.

WHILST examining aphonic patients with the laryngoscope we made several times the observation, that in those cases where nothing abnormal presented itself except an imperfect closure of the vocal cords, an im-

---

* Those who have made the application of galvanism to laryngeal disease a study, both at home and abroad, will no doubt be somewhat surprised, if not startled, at an assertion, based upon the doubtful evidence of one single case, which appeared lately in a paper entitled "The Medical Use of Electricity," originally published in the New York *Medical Record*, wherein it is broadly asserted that the constant current was applied with good effect to relieve the irritating effects of nitrate of silver, and that follicular inflammation of the entire mucous membrane of the pharynx and larynx of fifteen years' standing was cured by a few local applications of electricity. We are unfortunately not told where the electricity was applied.

provement speedily took place, by simply obliging the patient to pronounce certain syllables and words, during which action the laryngeal mirror was retained, when after a few attempts to speak the voice would materially improve.

Dr. von Brüns, who reports some very interesting results in this direction, has named this procedure the *gymnastic of the larynx.*

All patients who have been subjected to this treatment by Dr. von Brüns, and our observations coincide with his, had lost their voice at least several months before it, or had previously lost all resonance, and were reduced to hardly a whisper. Laryngoscopic examination disclosed no anomaly, except that the vocal cords retained during the act of speaking the same position and tension as during breathing. Sometimes a slight deviation and approach to each other was noticeable.

The course to be pursued is as follows: Whilst the laryngeal mirror is retained in the throat and the vocal cords are carefully watched, the patient is ordered to pronounce with great force and suddenly, simple vocal sounds and diphthongs, as: *æ, e, œ, ou.* At first this is very difficult, and instead of the plain vocal, a strong, blowing, piping and screeching sound is heard, which gradually is superseded by the proper sound. At this stage of the proceeding, already feeble motions of the vocal cords are observed. Next the vowels are to be given at varying pitch and at different length. If this is easily accomplished, the patient may then add to the long vowel one or more consonants, or pronounce monosyllables such as: *a-a-l, a-a-r, a-a-h, a-a-cht, au-au-s, au-au-f, i-i-u, ei-ei-n,* etc.* This accomplished, patients may now count from one to twelve, and try to pronounce polysyllables, words or sentences.

Those who have been vocalists, may be made to sing first single notes, changing constantly the pitch. At first the result is a crowing sound only, but when the notes become more clear, the patient is then requested to sing the scale, the third, fifth and eighth or octave, ascending and descending. During all this the vocal cords must be watched in the mirror, which is all this time retained in the throat. After this practice has been continued for some time and whilst the patient is engaged in singing the notes or pronouncing words, as the case may be, the mirror is then gradually removed without the patient being aware of it, first toward the median line of the mouth and finally entirely, and this practice is thereafter to be continued without the introduction of the mirror.

Curious phenomena come sometimes under observation during this practice. Thus von Brüns relates the case of a boy, who after three weeks of constant practice could sing in a beautiful soprano voice the scale and the third, fifth and eighth notes of the same, and yet about the same time was unable to pronounce a single word in a clear, distinct tone. Another boy who had already learned to sing a song and to articulate the words clearly, was unable to recite the words he sang distinctly—his effort was decidedly sporadic.

---

* As given by von Brüns, p. 341.

A similar case happened to us last winter, when after repeated gymnastic exercises on the larynx of a young lady troubled with hysteric aphonia some four months, she succeeded after several days to sing a little air, yet was unable to pronounce the words belonging to it distinctly. Ultimately she recovered.

This gymnastic must be practiced daily at least from fifteen minutes to half an hour. It may have to be continued for weeks, and in some cases may bring about the desired result after a few séances. We have as yet not been so fortunate as Dr. von Brüns to have cured a fair patient in one sitting.[*]

Case 1.—Mr. B——, æt. 30, tuberculous aphonia, the result of hæmopthysis, twelve weeks ago. Laryngoscopic examination reveals the vocal cords relaxed, pale, and the whole structure anæmic. Recovered after 9 séances, partially.

2.—Miss B——, æt. 19, vocalist, aphonic in consequence of a cold caught some three weeks previous, and too early and persistent attempt to sing immediately thereafter. Condition of the larynx normal, except epiglottis depressed. Recovered after 7 sittings.

3.—Miss A——, æt. 23, aphonia of eight months standing, the result of catarrh. Recovery slow after nearly three months treatment. No relapse.

Other cases were treated by laryngeal gymnastics, as well as by galvanism at the same time; the results were favorable, though varying in degree.

May it not be fairly presumed, that recovery from

---

[*] Opus p. 242.

aphonia of long standing after employing the means indicated, sometimes only for once, is owing to the same causes as recovery due to the influence exercised by sudden impressions upon the nervous system upon similar complaints sometimes?

Pseudo-Aphonia, simulated for some reason or other by individuals, such as are liable to military duty, or women anxious to be received into hospitals for simulated throat affections, can easily be brought to the test and the imposition discovered by laryngoscopic examination, when the pretended victim will readily express all sounds and words and the vocal cords respond to all requirements. Von Brüns thus exposed the imposition of three women who were anxious to be admitted into hospital although perfectly well.

CHAPTER VI.

REMOVAL OF FOREIGN BODIES FROM THE PHARYNX, LARYNX AND TRACHEA.

Morsels of food, small pieces of money, pins, needles, fish-bones, tooth-picks, artificial teeth, and numerous other substances, may, during the act of swallowing, laughing, sighing, screaming, running, or during deep inspiration, become suddenly lodged among other places in the fossæ of the root of the tongue, on either side of the epiglottis, in the ventricles of Morgagni, in the glottis or below the glottis in the trachea, causing there during the continuance of the lodgement great discomfort, sometimes inflammation and often death.

In anti-laryngoscopic times, most of these cases, at least those where the foreign body lodged in the glottis and below, unless voluntarily expelled by coughing or muscular contractions, proved fatal. But how changed is the condition of things to-day! The laryngoscope not only enables us to see the body in question, but also to seize and remove it.

Many successful cases are already reported. Thus, Gibb* in 1861, removed by aid of the laryngeal mirror, from a gentleman æt. 72, a pin which had lodged in the fissure formed by the pharyngeal wall at the outer, and the fold of the epiglottis on its inner side. This author relates a number of very interesting cases.

Mr. Paget† records an instance where a gold palate plate with nine artificial teeth was impaled in that same situation in a gentleman aged 60, for the long period of three months, and had actually lain out of sight for that period of time.

A similar case, which resulted fatally before aid could be rendered, is also reported in the *Lancet* of Nov. 20, 1862, page 591. Czermak‡ also mentions several cases of small bodies that have been extracted from the valleculæ, in which they had been imbedded.

Prof. Türck removed a fish-bone, 1¼ inches long, from the throat of a man æt. 64, which had lodged between the left corner of the tongue-bone and the free

---

* Opus cit. p. 404.
† Medical Times, Jan. 16, 1862.
‡ Wie. Med., 1865, Nov. 4.

margin of the epiglottis, at the first effort of removal. He also removed, in April 1866, a penny-piece (Kreuzerstück) from the sinus pyriformis of a juggler who had swallowed it during a performance and where it remained for two days.

We removed from the throat of a gentleman æt. 36, strong and stout, a piece of a tooth-pick, which he was chewing at table, and whilst suddenly laughing at some remark that was made, he swallowed, and the piece lodged in the left arytenoid fold its loose end pointing toward the free left border of the epiglottis. With the laryngeal mirror, the object could be seen distinctly, and was removed after the first trial with Türck's laryngeal pincers.

Mrs. M——, swallowed a needle, suffered great pain, and was brought to me some four hours after the accident had happened. Upon examination saw it implanted near the right arytenoid cartilage and firmly lodged in the epiglottis at the other end. First depressed the needle, having seized it with Türck's pincers, and then raised and removed it.

Mr. F——, swallowed a bristle from a tooth-brush in the morning, which kept annoying him till it became intolerable towards evening, when he consulted me. I found the bristle lodged in the right aryepiglottic fold, where I showed it to a friend of the patient. After considerable trouble the bristle was extracted with the pincers and Mr. F. felt instantaneously relieved.

140   THE PRINCIPLES AND PRACTICE OF

The following wood-cut represents Türck's pincers for the removal of foreign bodies.

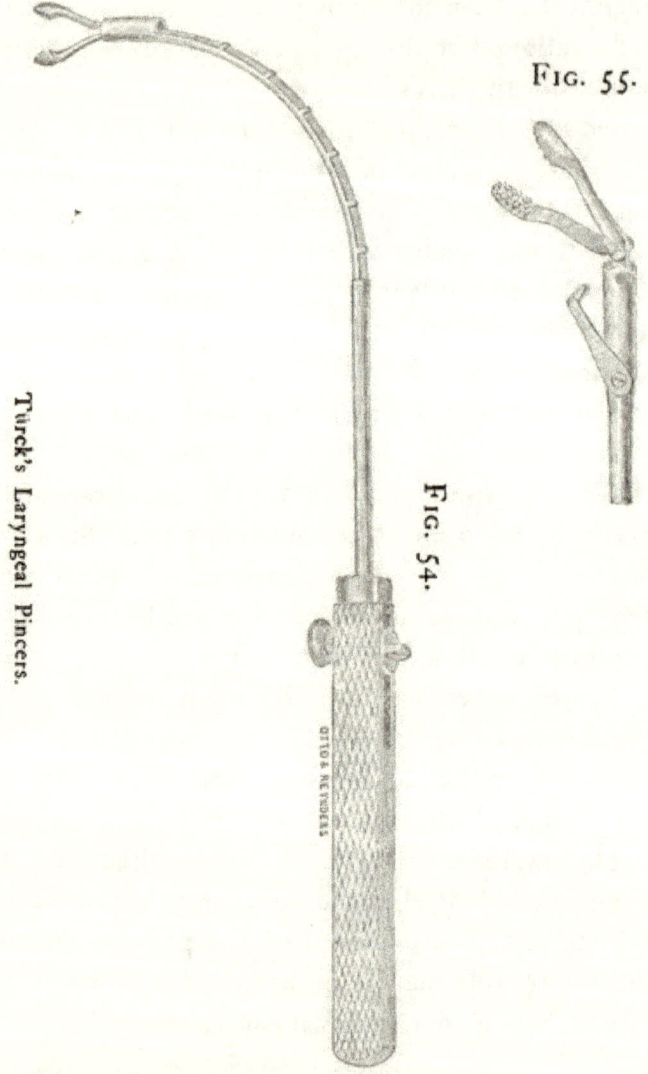

Fig. 54. Laryngeal pincers reduced.
Fig. 55. The same without handle, the blades separated; also the joint where they are joined.

CHAPTER VII.

ANÆSTHESIA DURING OPERATIONS IN THE LARYNX.

THE subject of local anæsthesia during laryngeal operations has engaged the attention of most laryngoscopists.

All seem however to agree that we do not as yet possess a reliable local anæsthetic, which might be so complete, as to enable us to execute lengthy operations without pain. We are therefore reduced to partial anæsthesia. Unsatisfactory as this is in most cases, yet in the absence of such an agent, for our purpose, at present, we have to content ourselves with such as will diminish the great sensibility of the organ and membranes covering the same.

A few inhalations of chloroform or æther are favorably spoken of by Gibb* and others.

Türck applies a solution to the larynx composed of morphia sulph. three grains, dissolved in one drachm concentrated acetic acid, the whole to be mixed with half an ounce of chloroform. We have applied this solution with success. Remes, Hartle and others, recommend bromide of potassium in doses of one or

---

* Opus, cit.

two grains; Gibb, the bromide of ammonium; Tobold, Lewin and von Brüns cold inhalations of tannin.*

The best—until there is discovered an anæsthetic to meet our purposes—of all agents to overcome irritability of the larynx and pharynx and to destroy great sensibility, is the constant introduction of the instruments for successive days into the throat of the patient, until they can be borne without causing any irritation at all.

In this manner the patient can attain great control over himself, so much so that it is even unnecessary sometimes to have his head held in proper position by assistants during the operation.

CHAPTER VIII.

CONCLUDING REMARKS.

CZERMAK stated somewhere, that many attempts at laryngoscopy and rhinoscopy had failed in the hands of students and physicians, because, discouraged by the failure of their first experiments in the art, they were disappointed and dropped the whole subject as unworthy of future investigation. This however is no more true in regard to our art than is illustrated daily in every difficult undertaking. He who lacks, therefore,

---

* Hypodermic injection has been tried by von Brüns to allay irritability of the larynx, but we have thus far no satisfactory data of the result.

that important requisite "*perseverance*," had better not attempt to walk along its crooked paths. But to him, who possesses that sine quâ non, will be opened a rich store-house of knowledge just beginning to be explored and the fruit ripe for gathering.

Many instruments have been described in this essay (we say this for the encouragement of beginners), that are not positively necessary to the simple practice of our art. A good light, a reflector and laryngeal mirror, with a few brushes, will suffice for first attempts, but he who desires to make laryngoscopy a study, let him not hesitate to get the best and ample instruments at once, regardless of expense, and disdain the idea of reducing everything connected with it to the "*cheap principle.*"* A man cannot work well in this or any other useful calling without good tools. When he has these, it is only after the eye and hand have been much practiced in applying remedies to the larynx, that instruments such as the forceps or lancet, or porte-caustique can be used with safety.

Time accomplishes these things.

Is there however no danger of too much medication being applied to the larynx? There is indeed danger of being too meddlesome, as well as not doing enough in this direction. For neither fault is there, however, now-a-days an excuse, since the laryngoscope has

---

* All the instruments mentioned in connection with the subject discussed are to be obtained of the well known and obliging firm of *Messrs. Otto & Reynders, No. 64 Chatham street, New York*, to whose ability and readiness to serve me, as I have no doubt they will other laryngoscopists, I bear most willing testimony.

brought these parts so completely within our reach that "to see" is almost "to do."

The most valuable additions to the diagnosis and treatment of the diseases of the larynx and trachea have been made since the discovery of the art of laryngoscopy and rhinoscopy. With its advent commenced a proud era in the history of practical medicine, from which we may reasonably expect still more flattering results in the future.

APPENDIX.

To page 56.—INFRA-GLOTTIC LARYNGOSCOPY OR TRACHEOSCOPY.

We append Fig. 56, the small mirror of Dr. Neudörfer, and also the fenestrated canula, into which, after tracheotomy has been performed and the canula properly adjusted, the mirror is introduced with its reflecting face directed upward and forward.

FIG. 56.

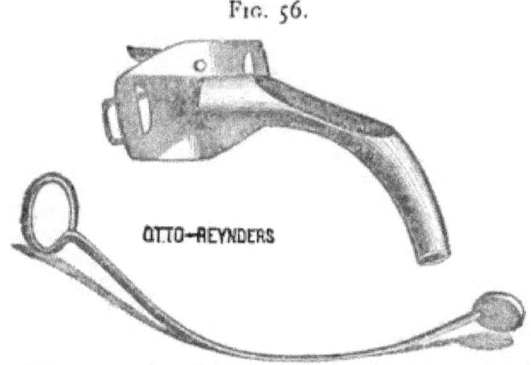

Mirror of Neudörfer, with Fenestrated Canula.

The canula must be provided with the largest possible window. The mirror must be made of thin plates of steel, about the size of a lentil, and must, of course, as in laryngoscopic examination, be warmed before introduction.

Czermak, to prevent the frequent interruption of the

examination caused by the rapid cooling of these little mirrors, had them covered with a very thin layer of dissolved caoutchouc, by means of which the image remains clear for a longer time.

The procedure is as tedious as it is difficult. In a healthy condition of the larynx, even, it must be remembered, that the lower surface of the vocal cords is not white, but reddened, resembling the mucous membrane of the larynx; hence they are recognized only by their motion and not by their color; the lower surface of the epiglottis may be seen, if the glottis is open. Thus Czermak claims to have thrown light into the larynx.

That the difficulties of examination are vastly increased when we have (as is almost always the case) to deal with pathological changes, is self-evident. The space within the larynx is not unfrequently narrowed in consequence of the swelling of the mucous membrane below the glottis, after the operation. But as Semeleder justly remarks,* the worth of the method lies in this, that precisely in pathological cases the epiglottis, either from tumefaction or from the shortening consequent upon the cicatrix, preserves so unfavorable a position, that the examination from above is very unproductive. In this way, also, by the aid of the mirror, the sound and other thin instruments might be introduced.

The same author has reported a remarkable case of laryngoscopic self-observation from below through the canula.

Czermak has also reported two observations.

---

* Dr. Caswell's translation, page 96.

APPENDIX.

The application of this instrument to the examination of the larynx must of course be very limited, yet it is exceedingly valuable and interesting. It cannot be urged too much upon those having an opportunity to examine the larynx from below, as directed above.

To page 79.—THE SYRINGE OR LARYNGEAL INJECTOR.

In addition to the syringes of Türck and Gibb, given on page 79, we present here Fig. 57, Tobold's Laryngeal Syringe, with mirror attached.

FIG. 57.

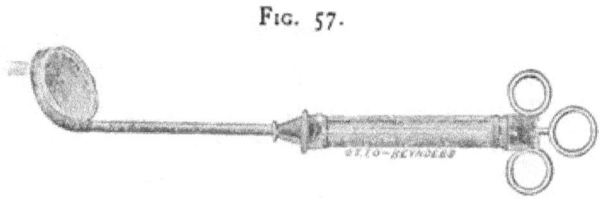

Tobold's Laryngeal Syringe, with Mirror.

This instrument will be found useful in cases where the operator desires to observe during the introduction of fluids into the larynx or trachea the direction the liquid takes, combining the laryngeal mirror with the syringe. The only objection is, that the mirror gets easily bespattered by the fluid injected, which may partially obstruct the view and therefore defeat the object intended.

Where the quantity of fluid to be injected is however small, there is nothing to be apprehended from this objection. The manner of holding the instrument will also contribute much toward the successful execution both of the operation and our purpose.

APPENDIX.

To page 86.—SPRAY-PRODUCERS OR PULVERIZERS.

To the various spray-producers mentioned page 86, et seq., must yet be added Schnitzler's Spray-Producer for the nose and pharynx, represented by Fig. 58.

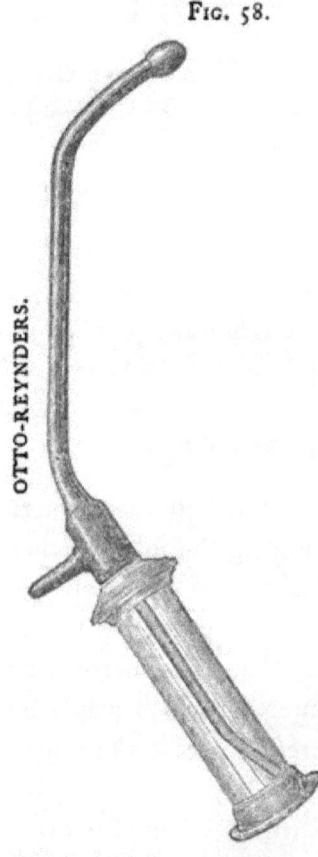

FIG. 58.

[Schnitzler's Spray-Producer.

It consists of a glass receptacle to receive the fluid to be pulverized, into the top of which is inserted a hard rubber tube, the smaller hollow extremity of which reaches almost to the bottom of the glass receiver. The free larger portion is of equal length but bent near its extremity at an obtuse angle and terminating into a bulb pierced with holes. A little above its insertion into the receiver is placed a short, hollow arm at right angles, into which the tube of Clark's hand-ball anatomizer is inserted. The surplus air introduced into the glass reservoir by means of steady pressure upon the lower ball causes the fluid to be expelled from the numerous holes in the bulb in the form of fine spray.

This instrument will be found principally useful in the treatment of catarrh.

APPENDIX. 149

To Page 99.—APPLICATION OF SOLID CAUSTIC.

Among the many useful instruments for the application of solid caustic to the larynx, one of the most complete is that of Professor Türck.

FIG. 59.

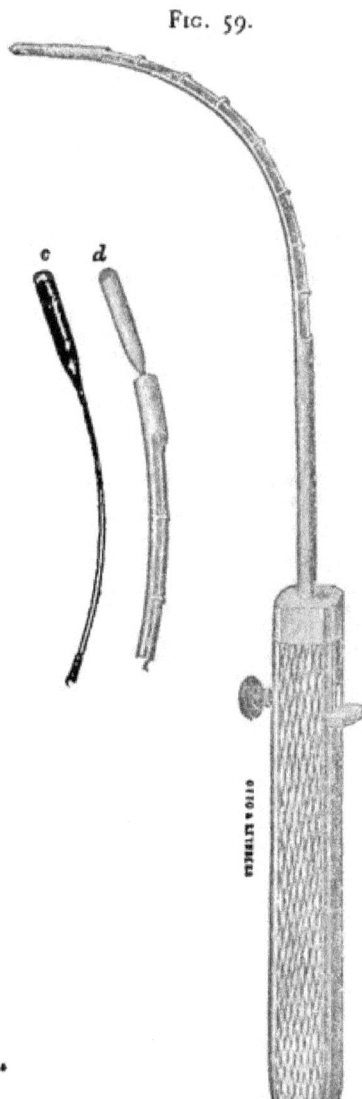

Türck's Porte-Caustique.

The large instrument is represented somewhat reduced in size; figs. *e* and *d* are of natural size.

It has already been stated, that the Porte-Caustique is not to be employed indiscriminately, but is almost indispensable in healing up obstinate ulcers, or to destroy growths that cannot otherwise be removed.

A patient presented himself two months ago, with an old obstinate syphilitic ulcer on the left aryepiglottic fold, which had resisted former treatment, although besides the use of internal remedies nitrate of silver in solution had been faithfully applied. Türck's Porte-Caustique being called into service six times in the course of two weeks, and the constitutional remedies being continued, the ulcer healed completely, to the great delight of the patient.

Magnifying Instruments, Micrometers, and Double Mirrors.

No mention has been made in its appropriate place of the use of magnifying instruments and micrometers in practical laryngoscopy, for the reason that both these instruments may safely be regarded as of no practical use in the treatment of disease.

The former, intended to increase the size of the laryngeal image, and recommended first by Dr. Wertheim of Vienna, in 1859, the latter constructed by Merkel of Leipzig, and Mandl of Paris, so as to measure the exact size of different parts of the larynx, and for estimating distances, both attest the ingenuity of the inventors, without being of use to the practitioner.

The same remark applies to the use of double mirrors, first suggested by Czermak, by means of which views of the floor of the nasal cavity are to be obtained. Dr. Wagner and Dr. Voltolini alone have thus far tried double mirrors. Wagner declares this method of experimenting as very laborious, the illumination does not always succeed, and the image which is obtained includes but little.

FINIS.

INDEX.

Abscess in the larynx, opening of, 103
    at the base of the epiglottis, case of, 108
Anæsthesia, of the larynx, 141
Aphonia, 126
    treated by galvanism, 128, 129, 130, 131, 132, 133
Apparatus for illumination, 21
    position of, 23
Applications of solutions to larynx, 67
Aryepiglottidean fold, 33
Arytenoid cartilages, 33
Asthma, cases treated by inhalation, 96
Auto-laryngoscopy, 42
Avery, Mr., experiments in laryngoscopy, 4

Babington, Dr., his glottiscope, 2
Baumés, Dr., his laryngoscope, 3
Bead, glass, in the left nostril, 58
Bennati, Dr., his laryngoscope, 3
Bozzini, Dr., his laryngoscope, 2
Bromide of potassium, 41
Brush, the laryngeal of Türck, 70

Cartilages, Santorini, 33
    of Wrisberg, 38
Cases, illustrating the use of the rhinoscope, 57
Catarrh, cases of, treated, 62, 72
    simple, by inhalation, 91
    chronic, by inhalation, 93
    of larynx and bronchi, 94
Caustic, solid, application of, 99

Cautery, electric, 122
Cauterizer, laryngeal, 99, 148
Color, healthy, of the larynx, 61
    of the turbinated bones, 62
    of orifices of Eustachian tubes, 62
Cricoid cartilage, 33
Czermak, his first labors, 7
    borrows Türck's mirrors, 7
    his claim to priority reviewed, 9
    his travels, 13
    introduced artificial light, 13
    demonstrated rhinoscopy, 49
    apparatus for auto-laryngoscopy, description of, 43
    first practice of tracheoscopy, 47
Changes in position of mirror, 28

Direct light, 18
Direct laryngoscopic examination, 26
Durham, Mr., case of cystic tumor, 110

Ecraseur, laryngeal, 121
Epiglottis, healthy, 33
    deviations from its natural position, 37
    illustrations of, 37
    position among mankind, 39
Epiglottic pincette, 38
Escharotics, 101
Essentials for laryngoscopic examination, 15

Face-shield, 88
False vocal cords, 33

INDEX.

Fauces, great sensitiveness and excitability, 40
Fish-bone removed from larynx with forceps, 138
Forceps, laryngeal, 140
Foreign bodies, removal of, 137
Fossa, innominata, 33

Galvanism, application of, to the larynx, 126
Galvano-cautery, 122
Galvanizer, laryngeal, 120
Garcia, Mr., his observations, 5
    his auto-laryngoscope, 42
Gibb, graduated laryngeal syringe, 79
    laryngeal fluid pulverizer, 86
Glottis, 33
Gymnastic of larynx in aphonia, 133

Head rest, 27
Healthy larynx, 35
History of the invention, 1
Hyoid fossa, 33
Hypertrophy of tonsils, 41

Johnson, Dr., his method of practicing auto-laryngoscopy, 42

Illumination, 18
Inhaling apparatus, Siegle-Bergson, 87
Infra-glottic laryngoscopy, 47
Injector, laryngeal, 86

Lamp, different kinds of, 19
    its position, 23
Lancet, laryngeal, 107.
Larynx, healthy, in natural position, 33
    as seen in the mirror, 33
Laryngeal mirror, 15
Lewin's pulverisateur, 84
Light, natural and artificial, 18
    reflected, 19

Magnifying instruments, 150
Mackenzie, laryngeal galvanizer, 129
Micrometers, 150
Mirrors, double, 150
    laryngeal, 15

Morbid growths, removal of, 111
    by cutting, 113
    by ligation, 121
Morgagni, ventricle of, 33
Morphine in Chloroform, 41
Moura, his pharyngoscope, 21

Neudörfer, Dr., his canula mirror, 145

Obstacles, encountered in laryngoscopic examination, 35
Obstruction of nasal passage, 57
Œdema, treated by scarification, case of, 108

Palate, hook, 50
    lasso, 52
Phthisis, tubercular, treated by inhalation, 97
Position of patient and physician, 26
    of head and tongue of patient, 27
    of hand and mirror, 30
Powders, the application of, to the larynx, 98
Practice of laryngoscopy, 15
Profile of laryngeal mirrors, 17
Prize Montyon, 1861, 12
Pseudo-Aphonia, 137
Pulverisateurs, 80

Recipro-Laryngoscopy, 45
Reflector, description of, 17
    position of, 17, 19
Reflected light, compared with direct, 18
Rhinal image, 54
Rhinoscopy, history of, 49
    by whom first practiced, 49
    how practiced, 53
    objects seen in examination, 54, 55
    cases illustrative of the practice of, 57

Scarification of the larynx, 103
Sellique, laryngeal speculum, 3
Senn, Dr., his laryngeal mirror, 2

INDEX.

Siegle, Dr., his experiments, 86
    his apparatus, 87
Smyly, Dr., his mode of demonstrating a patient's larynx to others, 45
Solutions, when applied, 67
    used, 68
    how applied, 69
Special difficulties in laryngoscopy, 35
Sunlight, 18
Syphilitic ulcers of larynx and pharynx, 95
Schnitzler, spray-producer, 148

Thyroid cartilage, 33
Tobold, Dr., his apparatus for illumination, 25
    phantom, 31
    laryngeal syringe, with mirror, 146
    polypus instruments for decision, excision, puncture, 114, 115
Tongue, the, 36
Tracheoscopy (see Infra-glottic laryngoscopy), 47
Trousseau, laryngeal speculum, 3
Turbinated bones, diseased, 59
Türck, Dr., revives the laryngoscope in medicine, 6
    first experiment, with laryngeal mirror, 6
    lends mirror to Czermak, 7
    just claim to priority, 9, 10, 11
    his laryngeal mirror, 16
    his reflector with spring, 20
    his independent apparatus for illumination, 22
    his tongue-spatula, 51
    his palate-lasso, 52
    his laryngeal brush, 70
    his sponge-carrier, 78
    his sponge-syringe, 79
    his laryngeal lancet, 107
    his polypus instruments, 116, 117, 118, 119, 120
    his pincette, 119
    his laryngeal pincers, 140
    his porte-caustique, 150

Ulcers, on the vomer and turbinated bones, 58

Ventricles of the larynx, 33
Vocal cords, 33

Warden, his prismatic laryngoscope, 4
Wertheim, his magnifying mirrors, 148
Wrisberg, cartilages of, 33

www.ingramcontent.com/pod-product-compliance
Lightning Source LLC
Chambersburg PA
CBHW031453160426
43195CB00010BB/962